T0226641

METHODS IN GENE TECHNOLOGY

Volume 2 • 1994

METHODS IN
GENE TECHNOLOGY

Editors: JEREMY W. DALE
PETER G. SANDERS
Molecular Microbiology Group
School of Biological Sciences
University of Surrey, Guildford

VOLUME 2 • 1994

 **JAI PRESS LTD**

London, England   *Greenwich, Connecticut*

JAI PRESS LTD
The Courtyard, 28 High Street
Hampton Hill, Middlesex TW12 1PD

JAI PRESS INC.
55 Old Post Road, No. 2
Greenwich, Connecticut 06836-1678

ISBN: 1-55938-264-3

A CIP record is available from the British Library for Volume 1
of this serial (Biotechnology, 660.605)

Printed and bound in the United Kingdom
Transferred to Digital Printing, 2011

CONTENTS

LIST OF CONTRIBUTORS

Juan C. Alonso

Max-Planck Institute for Molecular Genetics
Berlin, Germany

Alan D.T. Barrett

Department of Pathology
University of Texas
Galveston, Texas

Michael G. Burdon

Division of Cell and Molecular Biology
School of Biological and Medical Sciences
University of St. Andrews
St. Andrews, Scotland

Maria T. Corsetti

IST - Istituto Nazionale per la Ricerca sul Cancro
and Department of Biochemistry
Genoa University Medical School
Genoa, Italy

Giorgio Corte

IST - Istituto Nazionale per la Ricerca sul Cancro
and Department of Biochemistry
Genoa University Medical School
Genoa, Italy

Jeremy W. Dale

Molecular Microbiology Group
School of Biological Sciences
University of Surrey
Guildford, England

Andrew J. Davison   MRC Virology Unit, Institute of
                                 Virology
                             University of Glasgow
                             Glasgow, Scotland

Karl Drlica   Public Health Research Institute
                             New York, N.Y.

Wafa El-Adhami   Division of Biochemistry and
                                 Molecular Biology
                             Faculty of Science
                             Australian National University,
                             Canberra, Australia

Ing Swie Goping   Department of Biochemistry
                             McGill University
                             Montreal, Quebec, Canada

Tobin J. Hellyer   John L. McClellan Memorial
                                 Veterans' Hospital
                             University of Arkansas for Medical
                                 Sciences
                             Little Rock, Arkansas

Barbara Inglis   Division of Biochemistry and
                                 Molecular Biology
                             Faculty of Science
                             Australian National University
                             Canberra, Australia

Alan D. Jennings   Department of Pathology
                             University of Texas
                             Galveston, Texas

Pamela Johnstone   Molecular Microbiology Group
                             School of Biological Sciences
                             University of Surrey
                             Guildford, England

Haruyasu Kinashi   Department of Fermentation
                             Technology
                             Faculty of Engineering
                             Hiroshima University
                             Higashihiroshima, Japan

Angus I. Knight   Leatherhead Food Research
                             Association
                             Leatherhead, England

Toshihiko Koga   Department of Preventive
                             Dentistry
                             Faculty of Dentistry
                             Kyushu University
                             Fukuoka, Japan

Gareth Lloyd-Jones   Landcare Research
                             New Zealand Ltd.
                             Hamilton, New Zealand

Muhammad Malik   Public Health Research Institute
                             New York, N.Y.

John C. Maule   MRC Human Genetics Unit
                             University of Edinburgh
                             Edinburgh, Scotland

Elizabeth Norman   Molecular Microbiology Group
                             School of Biological Sciences
                             University of Surrey
                             Guildford, England

Nobuo Okahashi   Department of Oral Science
                             National Institute of Health
                             Tokyo, Japan

James D. Procunier   Research Station
                               Agriculture Canada, Winnipeg
                               Manitoba, Canada

Peter G. Sanders   Molecular Microbiology Group
                               School of Biological Sciences
                               University of Surrey
                               Guildford, England

Andrezej Sasiak   Institute of Biosciences and
                                   Technology
                               Department of Biochemistry and
                                   Biophysics
                               Texas A&M University
                               Houston, Texas

Gordon C. Shore   Department of Biochemistry
                               McGill University
                               Montreal, Quebec, Canada

S. V. Primal S. Silva   Veterinary Biologics and
                                   Biotechnology
                               Animal Health Division
                               Agriculture Canada
                               Ottawa, Ontario, Canada

Richard R. Sinden   Institute of Biosciences and
                                   Technology
                               Department of Biochemistry and
                                   Biophysics
                               Texas A&M University
                               Houston, Texas

Peter R Stewart   Division of Biochemistry and
                                   Molecular Biology
                               Faculty of Science
                               Australian National University
                               Canberra, Australia

Amalio Telenti   Institute for Medical Microbiology
                                   University of Berne
                                   Berne, Switzerland

Elizabeth A.R.Telford   MRC Virology Unit, Institute of
                                      Virology
                                   University of Glasgow
                                   Glasgow, Scotland

D. Alan Underhill   Department of Biochemistry
                                   McGill University
                                   Montreal, Quebec, Canada

Jian-Ying Wang   Public Health Research Institute
                                   New York, N.Y.

Janice E. Whitby   Rabies Unit
                                   Central Veterinary Laboratory
                                   Weybridge, England

Peter A. Williams   School of Biological Sciences
                                   University College of North Wales
                                   Bangor, Wales

John J. Willoughby   Sandoz Institute for Medical
                                      Research
                                   London, England

INTRODUCTION

Each volume in this series describes a selection of techniques that are related, at least loosely, by a common theme; within this framework are included a review of the most important basic procedures together with a selection of recent advances in techniques. By this means, each volume should be a manageable size, and contain a manageable amount of information, while at the same time providing enough detail for the relative beginner to have confidence in their ability to carry out the latest methods. Thus, the first volume centred on gene probes, and progressed from the principles of hybridisation via isotopic and non-isotopic labelling and detection methods through to gene amplification and RFLP analysis.

This volume follows the same approach, applied broadly to the purification and characterization of nucleic acids. Electrophoretic techniques play a major role in both connections, including conventional techniques for the analysis and preparation of DNA, and also pulsed field techniques (in various manifestations) that have enabled electrophoresis to be applied to the isolation and analysis of larger fragments. Electrophoresis can be used to analyse the conformation of DNA as well as fragment size, as witnessed here by the powerful technique of single strand conformational polymorphism for analysing variation in specific DNA fragments, and the analysis of DNA supercoiling.

The random amplified polymorphic DNA (RAPD) technique provides an invaluable extension to conventional RFLP methodology for genomic fingerprinting, while techniques such as PCR walking are useful in isolating and characterising unknown sequences adjacent to a previously characterised sequence. The detection of DNA binding proteins, and characterising the sequences to which they bind, has become of major importance; gel retardation assays and DNA footprinting are widely used, while the SouthWestern blot technique offers potential that has perhaps not yet been fully realised.

Finally, DNA characterisation must include sequencing. A full coverage of this topic would require a separate volume (which may come later!), but in these days of automated sequencers we could not resist the temptation to include a chapter to demonstrate that it is possible to produce relatively large amounts of sequence data in other ways.

Safety

Many of the procedures described in this book use chemical reagents that are hazardous (toxic, carcinogenic or flammable); the use of these chemicals may be subject to local safety regulations. Similarly, the use of radioactive isotopes is intrinsically hazardous and is also subject to safety regulations.

Many users of electrophoresis equipment are not sufficiently aware of the hazards associated even with low voltages. All electrophoresis equipment should be designed and used in a manner that protects the user.

It is important that you should be aware of the potential risk arising from any of these procedures, and of the local regulations governing such hazards, including the safe use, storage and disposal of hazardous chemicals.

Jeremy W. Dale
Peter G. Sanders
Series Editors

Chapter 1

PURIFICATION OF DNA

Jeremy W. Dale

OUTLINE

Methods in Gene Technology, Volume 2, pages 1–13
Copyright © 1994 JAI Press Ltd
All rights of reproduction in any form reserved.
ISBN: 1-55938-264-3

1. INTRODUCTION

The purpose of this chapter is to provide some of the background to the commonly used general methods for the purification of DNA, that is (primarily) its separation from other constituents of a cell extract rather than the separation of different species of DNA molecules. These methods often rely on the physical chemistry of DNA, and especially of double stranded DNA. Although a full treatment of this topic is beyond the scope of this book, some misconceptions need to be cleared up.

The major misconception is that the two DNA strands are held together solely by hydrogen bonding between the bases. Although hydrogen bonding has a role to play, this ignores other effects, primarily those due to the hydrophobic nature of the bases themselves. This influences the structure in two ways: first of all, there is the stacking of the bases on a single strand by hydrophobic interactions between adjacent bases. This is facilitated by the helical structure which brings adjacent bases closer together. (Intercalating dyes such as ethidium bromide expand the structure by insertion between adjacent bases, thus reducing the need for twisting of the helix). Secondly, hydrophobic interactions between bases on opposing strands stabilise the double stranded structure, by reducing contact with water molecules.

The importance of the hydrogen bonding is to impose specificity on these interactions; in this context it is not so much the holding together of the two strands that is important, but rather that if the correct hydrogen bonding does *not* form then the bases will not be able to stack together correctly.

An additional consideration is the effect of the negative charge on the phosphate groups in the DNA 'backbone', which results in electrostatic repulsion between the chains. The addition of sodium chloride (or other salts) will tend to stabilise the double stranded structure since the sodium ions will counteract the negative charge on the phosphate groups. This is at least partly responsible for the apparently paradoxical effect that NaCl addition increases the stability of the double stranded structure, when superficially the opposite might be expected, by interference with hydrogen bonding. The cloud of positive counterions neutralises the negative charges, while the consequences of any interference with hydrogen bonding are comparatively small.

It is also necessary to consider the effect of the solvent on these interactions. For example, water has a high dielectric constant, which reduces the electrostatic force between two ions of opposite charge such as phosphate and Na^+. The addition of an organic solvent such as ethanol reduces the dielectric constant; in the presence of positively charged counterions, this will favour the formation of ion pairs resulting in insolubility of the DNA.

This is necessarily a superficial account of a very complex story: the effects of solutes and other solvents on water-water interactions also need to be taken into account, for example. Further information can be found in [1].

2.   PURIFICATION PROCEDURES

A crude DNA preparation, when extracted from the cell, will be contaminated with a variety of high molecular weight material such as proteins, RNA, and polysaccharides, as well as a broad spectrum of low molecular weight material. For many purposes, only the briefest of purification is necessary (or even no purification at all); although inactivation of nucleases is needed for most procedures, this can often be accomplished sufficiently by the inclusion of EDTA in the lysis buffer, or by heating the cell extract. For example, the preparation of template DNA for PCR amplification of a bacterial gene may require nothing more than boiling a dilute cell suspension, thus achieving both lysis and inactivation of nucleases simultaneously.

The extreme sensitivity of PCR means that only a very low DNA concentration is required, thus reducing the problem arising from contaminating material by allowing dilution of the extract. (Many problems with inhibitory material in DNA extracts used for PCR can be overcome by using less template.) However, for other procedures, higher DNA concentrations are needed. Contamination then becomes a problem, for example by interfering with restriction endonuclease digestion. Purification, and concentration, of the DNA is then necessary.

The purification and concentration procedures described below are also necessary for the removal of ingredients that are deliberately added to DNA during manipulation, but which will interfere with subsequent steps. Restriction endonucleases can often be inactivated by heat (but not always, and not necessarily reliably), but other enzymes (such as calf intestinal phosphatase) are less easily inactivated and need to be removed by phenol extraction. Low molecular weight ingredients (or by-products) from one step (such as buffers, ATP, phosphate, EDTA) may interfere with subsequent steps, and have to removed, for example by ethanol precipitation. However, it is important to realise that each purification step may result in either loss of, or damage to, your DNA, so it is a good policy to minimise the use of such procedures. This can sometimes be achieved by good experimental design, e.g. examining the order in which the various steps are executed, and considering whether intermediate purification is really necessary. For example many restriction enzymes will work adequately in sub-optimal buffers (but beware of partial digests and star activity); alkaline phosphatases will function well enough in most restriction buffers. Many of these procedures are also suitable for the purification of RNA; however, fuller information on RNA purification is provided in Chapter 9.

2.1   Removal of Proteins

2.1.1   Materials

Phenol

Use ultrapure (molecular biology grade) phenol, stored at −20°C. Melt the phenol at 68°C, and transfer an aliquot to a fresh container. Return the remainder to the freezer. Equilibrate the melted phenol with at least an equal volume of a suitable buffer (e.g. 100 mM Tris-HCl pH 8); repeat this step several times, until the pH of the aqueous phase is greater than 7.6. Store the equilibrated phenol, in contact with the equilibration buffer, at 4 °C. In use, be careful to take only the lower (phenol) layer. 8-hydroxyquinoline can be added to the phenol (final concentration 0.1%) as an antioxidant.

> *CAUTION*: phenol is highly corrosive and is toxic via skin absorption. *Always* wear gloves and eye protection when handling phenol, even in capped tubes. In the event of spillage, wash contaminated skin *immediately* with soap and water.

Chloroform and chloroform/isoamylalcohol

Although many procedures specify a chloroform/isoamylalcohol (24:1) mixture, in general chloroform itself performs quite adequately. In this chapter, where chloroform is specified, either can be used.

Note that special procedures (according to your local regulations) are necessary for disposal of chloroform; it should not be poured down the sink, even in small quantities.

Ether

Ether can (if absolutely necessary) be used to extract traces of phenol or chloroform from the DNA solution. It should be saturated with water before use, to avoid extracting water from the sample.

> *CAUTION:* ether is both volatile and highly inflammable; it must be used, and stored, in a suitable fume hood. Ether (or any material contaminated with ether) must *never* be stored in refrigerators or freezers as the vapour can accumulate to an explosive mixture. This is a major cause of laboratory fires.
>
> My recommendation is to avoid the use of ether altogether. There are very few occasions when its use is necessary.

TE buffer
    10 mM Tris–HCl, pH 8.0
    1 mM EDTA

Proteinase K
    20 mg/mL solution in water
    Store at –20°C

Proteinase K reaction buffer
    10 mM Tris–HCl pH 7.8
    5 mM EDTA
    0.5% SDS

2.1.2 Phenol extraction

1. Add, to the DNA sample, an equal volume of phenol (or a 1 : 1 mixture of phenol and chloroform). Mix by vortexing until a

milky emulsion is formed. Note: for larger DNA molecules, or in other situations where shearing is likely to be a problem, gentler mixing is indicated.)

2. Centrifuge briefly to separate the layers. The extent of centrifugation necessary will depend on the thoroughness of the mixing in step 1; experience is the best guide. If the layers are not completely separated, repeat the centrifugation. Precipitated protein may be visible at the interface.

3. Transfer the upper (aqueous) phase to a fresh tube, taking care to avoid the interface and lower (phenol) layer.

   Notes:

   (a) For maximum yield, and especially if there is a heavy protein precipitate, it is possible to back-extract the phenol layer: add an equal volume of TE buffer, mix thoroughly and centrifuge. Recover the aqueous (upper) layer and combine with the first aqueous phase. However, this dilutes the DNA, and is not usually necessary.

   (b) The above version of step 3 is the standard procedure. My preference, especially if treating a large number of samples, is to use a narrow pipette tip to remove (and discard) the lower (phenol) layer. This is quicker to do, reduces the number of tubes needed, and achieves better separation of the phases (due to the shape of the microfuge tube). If the possibility of a small amount of cross-contamination is not a critical consideration, the same pipette tip can be used to remove the phenol layer from all the tubes.

   (c) If the DNA sample contains a high salt concentration (e.g. in material recovered from a CsCl gradient), it may be more dense than the phenol, and the position of the two layers is therefore reversed, i.e., the lower layer may be the aqueous phase containing the DNA. If in doubt, add a small extra volume of phenol and observe which layer increases in size. If the phenol contains hydroxyquinoline, it will be yellow and can therefore be readily identified.

4. For more complete protein removal, and especially if there is a heavy protein precipitate, repeat steps 1–3.

5. Since phenol is to some extent soluble in water, there will be some phenol dissolved in the aqueous layer. This can be removed by addition of an equal volume of chloroform, mixing thoroughly and centrifuging briefly to separate the layers.

6. To remove traces of phenol and/or chloroform, chill the tube on ice, centrifuge briefly (preferably in a cold room), and remove the small amount of phenol or chloroform from the bottom of the tube. Alternatively, water-saturated ether can be used to remove traces of phenol and/or chloroform (but see the note in Section 2.1.1).
7. Recover the DNA by ethanol precipitation (see Section 2.3).

2.1.3 Proteinase K Treatment

Phenol (or phenol-chloroform) treatment is a very effective, and rapid, method for denaturing proteins, and is particularly useful for rapidly inactivating nucleases or other enzymes that may interfere with DNA manipulations. However, it does not necessarily *remove* proteins that are firmly attached to the DNA; this is especially relevant for example in the purification of linear plasmids (Chapter 13) which have covalently attached proteins. Phenol extraction, in the absence of treatment with a proteolytic enzyme, will leave denatured protein attached to the DNA.

Proteinase K, which is commonly used for this purpose, is a broad-spectrum serine protease. Its activity can be inhibited at the end of the reaction by the addition of PMSF (a serine protease inhibitor), but it is generally safer to use phenol extraction following protease digestion, in order to ensure the complete absence of remaining proteolytic activity.

'Pronase', a proprietary name for a non-specific protease from *Streptomyces griseus,* is often used as a cheaper alternative to proteinase K. Preparations of Pronase can contain detectable nucleases, and it has been the usual practice to preincubate (2 h at 37°C) stock solutions of Pronase to allow self-digestion to occur (i.e., the Pronase digests the nucleases). However, it is now possible to obtain commercial Pronase preparations that are nuclease-free.

Proteinase K digestion is carried out by addition of the enzyme (50 µg/mL) to the DNA sample, and incubating for 1 h at 37°C. Ideally, the sample should be dissolved in proteinase K reaction buffer, but sufficient activity will occur in many other buffers, especially with the addition of SDS to a final concentration of 0.5%. The presence of SDS partially denatures many proteins (without affecting the enzymic activity of proteinase K); proteolytic enzymes in general are much more active against denatured proteins than against proteins in their native state.

2.2   Removal of Other Macromolecules

The only other macromolecule that is easy to remove by specific treatment is RNA, which can be readily removed by treatment with ribonuclease (RNase). Traditionally, RNase preparations are heat-treated before use to destroy contaminating DNase activity: RNase is a notoriously heat-stable enzyme, while DNases are in general quite heat-labile. A stock solution of RNase (10 mg/mL in water) is boiled for 15 min and allowed to cool to room temperature; heating to a lower temperature (such as 70°C) is often used, but does not always seem to destroy DNases effectively. To remove RNA from your DNA sample, add RNase to a final concentration of 50 μg/mL and incubate for 1 h.

It is now possible to buy RNase that is DNase-free. However, the custom of heat-treating RNase solutions still persists, and it seems to do no harm.

A further consideration is at what stage to use RNase. Many protocols call for RNase to be added at an early point in the DNA isolation procedure. My preference (in for example, plasmid purification) is to leave the RNA in the sample for as long as possible, on the grounds that it results in increased yields of DNA, as it acts as a carrier in ethanol precipitation, and also reduces the adsorption of DNA to surfaces in the intermediate steps.

It is possible therefore to add RNase to the buffer in the final step of redissolving the DNA after ethanol precipitation—or even to add RNase to the restriction digest. In fact, if you're only interested in relatively large DNA fragments on an agarose gel (say greater than 2 kb), there is no need to remove the RNA at all.

Note: RNase is a very persistent enzyme and will contaminate pipettes, gel tanks, glassware, and anything else it comes into contact with. Therefore, any apparatus that is to be used for RNA work must be kept strictly segregated from activities with DNA.

Cell extracts may also contain other macromolecules, including polysaccharides and lipids, that can interfere with restriction digests and other manipulations. Since these may be bound firmly to the DNA, they are not easy to remove by physical means; although enzyme treatment may help, the variety of possible contaminants makes success unpredictable. In general, the best strategy is to avoid such contamination wherever possible, e.g. by appropriate selection of growth conditions. So if your digests do not work, and phenol extraction/ethanol precipitation does not improve matters, discard your DNA preparation and start again.

2.3   Removal of Low Molecular Weight Contaminants

Low molecular weight contaminating material can be removed by dialysis, provided that the dialysis tubing is carefully prepared (by boiling in 1 mM EDTA) and that the complete operation is carried out aseptically and without introduction of nucleases. Since this makes it rather a cumbersome process, other procedures are usually adopted. The most commonly used is ethanol precipitation, which also has the benefit of concentrating the DNA; a useful alternative in some circumstances is gel filtration.

2.3.1   Materials

Ethanol precipitation requires the presence of salts, which can be provided using one of the following stock solutions:
3 M sodium acetate, pH 4.8
5 M sodium chloride
7.5 M ammonium acetate

ethanol (or isopropanol)

70% ethanol

2.3.2   Ethanol Precipitation

1. Assuming the DNA is dissolved in a low ionic strength buffer such as TE, increase the salt concentration by adding 0.1 vol of sodium acetate. Other salts can be substituted: ammonium acetate (0.5 vol) has the advantage of volatility, so it is easy to remove. Sodium chloride (0.01 vol) is often used, but seems to have no advantages, and plenty of disadvantages, such as a tendency to precipitate out.

   If your DNA sample is already in a high salt buffer, there is no need to add more salt.
2. Add 2 volumes of ice-cold ethanol, and mix well (but not too vigorously; do not vortex).

   It is possible to use isopropanol rather than ethanol, in which case reduce the volume added to an amount equal to that of the

DNA sample.

3. Put the tube at −20°C (or −70°C if available) for 30–60 min. This can be extended indefinitely if required.

   Note: there is much disagreement as to whether this step is necessary at all. The really important factor is the prevention of a rise in temperature in the next step (centrifugation).

4. Spin down the precipitate in a microcentrifuge, at full speed, for 3–30 min. It is preferable to do this step in a cold room, if possible; for centrifugation times longer than a few minutes a cold room is essential.

   The time needed to spin down the DNA will be influenced by the concentration of the DNA and the size of the fragments. Since smaller fragments require longer centrifugation times (or higher speeds), it is possible to use ethanol precipitation as a crude method of size fractionation—for example in separating a PCR product (of say 500 bp or more) from unincorporated primers. This can be done by adding 0.2 vol of 10 M ammonium acetate and 1 vol of isopropanol and centrifuging immediately from 20 min at room temperature. A few trial experiments might be needed to get precisely the right conditions in each case.

   It is even possible to use selective partial ethanol precipitation for a crude size fractionation of restriction digests, as required for library construction. In this case, it is easier to vary the ethanol concentration: i.e. start by adding 0.5 vol of ethanol; spin, recover the supernatant (retaining the pellet), and add a further 0.5 vol of ethanol. Repeat until 2.5 vol have been used, then test each of the pellets for the presence of the appropriate size range of DNA fragments.

5. Using a fine pipette tip, remove *all* the liquid from the tube.

   Note that the DNA pellet will probably not be visible, but it will be on the wall of the tube towards the bottom (assuming use of an angle rotor); the pipette tip can therefore be safely taken right down to the base of the tube. Observe the tube as the liquid level falls: the DNA 'pellet' may become visible as a scanty, granular deposit on the wall of the tube—or its presence may merely be indicated by the liquid clinging to the tube. The plastic tube is hydrophobic and is not wetted by the liquid, except in the presence of the hydrophilic DNA deposit.

If you have a heavier precipitate than that, especially if it is white and/or at the base of the tube rather than on the wall— then you have a lot of contamination.

6. Add sufficient 70% ethanol (ice-cold) to cover the 'pellet' (half-filling the tube is adequate). Without resuspending the pellet, spin the tube again, keeping it in the original orientation in the centrifuge head.

   The purpose of this step is to wash out any remaining salts in the DNA pellet. If the DNA is resuspended in the 70% ethanol, it can be difficult to spin it down again. However, if there is an unusually heavy precipitate in step 5, then it will be necessary to gently resuspend the pellet in the 70% ethanol.

7. Carefully remove *all* the liquid from the tube. If necessary, spin the tube again to collect any remaining liquid in the bottom of the tube.

8. Dry the pellet *for a few minutes only*, in a vacuum desiccator.

   This will remove the last traces of ethanol; only a very brief treatment is necessary if ethanol removal at step 7 has been done thoroughly. Do *not* assume that it does not matter if some ethanol remains in step 7 as it will be evaporated anyway. Evaporation removes the ethanol but leaves salts behind.

   Extending the evaporation step for longer than a few minutes will remove too much water from the sample; over-dried DNA can be very difficult to redissolve.

9. Redissolve the 'pellet' in a small volume of water or TE buffer.

   Care is needed at this stage, as the pellet may not be visible and the volume of buffer will be too small to cover it. Use a pipette to spray the walls of the tube with buffer, repeatedly.

   Note: If the amount of DNA is small, the yield can be improved by using siliconised tubes and a carrier: tRNA (10 µg/mL) works well but can interfere with subsequent procedures. Glycogen (1–2 µg/mL) is a better alternative.

2.3.3   Gel Filtration

Gel filtration can be used for simple desalting of DNA, as an alternative to ethanol precipitation. With the use of spun columns, this is quicker than precipitation, but it is considerably more expensive and unlike ethanol precipitation does not concentrate the DNA. Gel filtration can also be used for separating DNA from RNA, size fractionating DNA fragments, and plasmid purification. See Chapter 5 for more information on gel filtration, and other column methods.

3. DETERMINATION OF DNA CONCENTRATION

Having purified the DNA, it is necessary to assess the quantity (and quality) of the product.

The simplest method of assessing the amount of DNA is to measure the absorbance at 260 nm. As a rough approximation, an absorbance of 1 corresponds to 50 µg/mL of double stranded DNA. For single-stranded DNA and for RNA, an A_{260} value of 1 indicates 40 µg/mL.

These values are only approximations, and become progressively less reliable with shorter oligonucleotides, where the absorbance depends on both length and base composition. A reliable determination of concentration should be based on the extinction coefficients of the individual bases [2], which are:

$$dGTP = 11.7 \times 10^3 \, M^{-1}$$
$$dCTP = 7.3 \times 10^3 \, M^{-1}$$
$$dATP = 15.4 \times 10^3 \, M^{-1}$$
$$dTTP = 8.8 \times 10^3 \, M^{-1}$$

The extinction coefficient ε of the oligonucleotide is obtained by adding the value for each base at each position. The molar concentration C can then be determined using the formula

$$C = A_{260}/\varepsilon \, M$$

If an approximate concentration is sufficient, this can obtained using the assumption that an A_{260} value of 1 corresponds to 33 µg mL^{-1} [2].

The above values are only useful if the absorption is solely due to the required nucleic acid. If there is contamination by other ultraviolet absorbing material (e.g. proteins), then the values will be incorrect. To guard against this, measure the absorption at 280 nm as well. Nucleic acids have an absorption maximum at about 260 nm, while proteins absorb more strongly at 280 nm. A ratio of A_{260}/A_{280} of less than 1.8 will indicate protein contamination. An A_{260}/A_{280} value greater than 1.8 does not mean the DNA preparation is exceptionally pure; pure

DNA will have a value of 1.8. A DNA preparation with a value greater than that is probably contaminated by RNA, which has a higher A_{260}/A_{280} value.

It is essential to appreciate that this measurement tells you nothing about the quality of your DNA, which may be extensively (or even completely) degraded and yet give a high A_{260} value.

The quality of the DNA is best assessed by running a sample on an agarose gel; if a known amount of a molecular weight standard is run on the same gel, a comparison of band intensity will give a reasonably reliable estimate of the amount of DNA—certainly as reliable as the spectrophotometric method.

REFERENCES

[1]   Calladine, C.R. and Drew, H.R. *Understanding DNA*. Academic Press, London, 1992.
[2]   Sambrook, J., Fritsch, E.F. and Maniatis, T. *Molecular Cloning: a Laboratory Manual*, 2nd edn. Cold Spring Harbor Laboratory Press, Cold Spring Harbor, 1989.

Chapter 2

ELECTROPHORETIC SEPARATION OF NUCLEIC ACIDS

Peter G. Sanders

OUTLINE

Methods in Gene Technology, Volume 2, pages 15–41
Copyright © 1994 JAI Press Ltd
All rights of reproduction in any form reserved.
ISBN: 1-55938-264-3

1.   INTRODUCTION

Gel electrophoresis is a simple yet powerful technique which enables DNA and RNA molecules to be separated according to their size and subsequently visualised by a number of techniques. This chapter is designed to provide protocols for a number of basic techniques in gel electrophoresis of nucleic acids so that those new to the field can successfully separate and analyse a variety of DNA and RNA molecules. A number of excellent laboratory manuals [1,2,3] cover this field in more detail and provide many additional protocols, and the reader should also consult these. For those concerned with the theory behind gel electrophoresis there are many recent reviews [4,5], and articles can be found in the series *Advances in Electrophoresis* and the journal *Electrophoresis*, both published by VCH Verlagsgesellschaft mbH.

2.   AGAROSE AND POLYACRYLAMIDE GELS

Molecules of DNA and RNA can be separated by gel electrophoresis using agarose, polyacrylamide or a combination of these in one matrix under native or denaturing conditions, the primary determinant of mobility being the size of the DNA. This chapter will consider the use of these matrices for electrophoresis in slab gels, and not in capillary systems which are becoming fashionable for certain separations. Agarose, a linear carbohydrate polymer extracted from a number of red seaweeds, is routinely used to produce a supporting gel matrix which can separate DNA fragments from tens of nucleotides in size to megabases, depending on the concentration of the agarose and the electrophoresis regime used. Separation of large DNA fragments by a variety of pulsed field gel electrophoresis (PFGE) regimes is the subject of Chapters 11–13 in this book.

Agarose is also routinely used to separate RNA molecules, usually under denaturing conditions which remove secondary structures that can affect the mobility of the RNA through native gels.

Many grades of agarose can now be bought from reagent suppliers, including specially purified and modified agaroses with reduced levels of enzyme inhibitors, which melt and gel at lower temperatures (65°C and 35°C respectively), and which have lower levels of background fluorescence after staining with ethidium bromide (ETBR). Agaroses with lower gelling temperatures are useful for preparative agarose gel electrophoresis (see Chapters 3, 7 and 13), but the choice of grade is important to ensure no inhibitors of subsequent reactions are present.

Preparation of an agarose gel matrix for the electrophoretic separation of nucleic acids is extremely simple, agarose powder is dissolved by heating in a suitable buffer solution and, after cooling, poured into an appropriate mould or gel former. The concentration of the agarose in the gel determines the sizes of the pores through which the DNA must travel, the more concentrated the gel the smaller the fragments that can be resolved (Table 1).

Polyacrylamide gels are more complicated to produce and are made by polymerising acrylamide with the cross linker N,N'-methylenebisacrylamide (usually mixed at a ratio of 30 : 1) in the presence of free radicals provided by ammonium persulphate, and the catalyst and stabilising reagent, TEMED (N,N,N,N'tetra-methylenediamine). Oxygen must be kept away from the polymerising acrylamide.

The percentage of acrylamide in the gel determines the length of the chains of polymerised acrylamide and this with the crosslinking provided by the methylenebisacrylamide determines the sizes of the pores through which the DNA must pass. The higher the concentration of acrylamide the smaller the pores and the smaller the fragments of DNA that can be separated (Table 1). For different percentage gels the marker dyes, usually bromophenol blue and xylene cyanol FF allow the migration of double stranded DNA fragments to be monitored. Bromophenol blue migrates with double

Table 1.   (a) Separation of DNA fragments in (a) agarose gels [2,6]
             and (b) polyacrylamide gels [2,6].

(a) Agarose gels

Fragment size (kb)	Percentage of agarose (w/v)
60–5.0	0.3
20–1.0	0.6
10–0.8	0.7
7–0.5	0.9
6–0.4	1.2
3–0.2	1.5
2–0.1	2.0
1–0.01	4.0[a]

[a]including composite gels

(b) Polyacrylamide gels

Fragment size (bp)	Percentage of polyacrylamide (w/v)	Xylene cyanol	Bromophenol blue
2000–1000	3.5	460	100
500–80	5.0	260	65
400–60	8.0	160	45
200–40	12.0	70	20
150–25	15.0	60	15
100–6	20.0	45	12

stranded DNA of about 300 bp in agarose gels and 100 bp in 3.5%
polyacrylamide gels, and xylene cyanol FF migrates at 4000 bp in
agarose and 460 bp in polyacrylamide gels (2,6).

Polyacrylamide gels have traditionally been used for the separation
of smaller DNA fragments (1000 bp and smaller) which may differ in
size by only one or two base pairs. They have been invaluable for
separating single stranded molecules differing in length by one base
in DNA sequencing programmes (see Chapter 10). DNA eluted from
polyacrylamide gels is normally purer than that prepared from agarose
which can contain inhibitors that prevent subsequent enzyme
reactions from working efficiently, although the elution of DNA by
ion exchange columns and glass milk regimes can usually remove
these impurities.

Agarose gels are normally prepared as horizontal slab gels on a supporting platform, whereas polyacrylamide gels are prepared as a sandwich between two glass plates which must be tightly bound against the spacers and sealed at the bottom. Electrophoresis of polyacrylamide gels takes place in the vertical position. Oxygen prevents polymerisation of the acrylamide, precluding the routine use of horizontal gels, and insertion of the comb between the glass plates when forming the wells protects most of the acrylamide at the top from oxygen exposure. Only the most exposed areas of the acrylamide therefore normally fail to polymerize.

For the separation of very small fragments of DNA polyacrylamide gels have traditionally been used but high percentage agarose gels, particularly containing mixtures of agaroses, are often now used as substitutes for polyacrylamide and can separate fragments down to 10 bp [7] (see later). For preparative electrophoresis in our laboratory agarose has become the routine medium and many commercial organisations now provide reagents (e.g. glass milk, as Gene Clean, Stratech Ltd; mini ion exchange columns, Quiagen Tip-20, Hybaid) which rapidly and efficiently purify DNA from agarose, and examples of such techniques can be found in Chapter 5.

3. FACTORS AFFECTING NUCLEIC ACID MOBILITY IN GELS

The migration of DNA through a gel matrix is affected by a number of physical parameters including the size and conformation of the DNA, the agarose or polyacrylamide concentration, the presence/absence of ethidium bromide (ETBR) and the applied electric field.

The base composition of the DNA has little effect for electrophoresis through agarose but anomalous migration of many fragments does occur through polyacrylamide, preventing the use of this medium for accurate sizing of DNA fragments [8,9].

The effect of the DNA's size and conformation on its mobility is not surprising as the gel matrix is an interconnecting network of pores, which decrease in size as the gel concentration is raised. The concentration of agarose in the matrix therefore also affects nucleic acid mobility and there is a linear relationship between the logarithm of the DNA's electrophoretic mobility and the concentration of the gel. As shown in Table 1, by changing the concentration of the agarose in a gel a wide range of sizes of DNA molecules can be separated.

When an electric current is applied, linear double-stranded DNA molecules are considered to travel through these pores by reptation, a worm-like end-on migration (see Chapter 11), [10–12] although other mechanisms of migration through pores have been postulated [13] with some support [14]. The rate of travel of linear double stranded DNA is inversely proportional to the $\log_{10}$ of the size in base pairs (bp) and this provides a simple way of sizing fragments of unknown size if electrophoresed through the most appropriate concentration of agarose [15]. Measuring the distance travelled by a fragment of unknown size allows comparison to the distance migrated by marker DNA of known size in bp. The marker digest is plotted with $\log_{10}$ bp as the Y axis and distance migrated as the X axis. Plotting the data on linear–logarithmic (semi-log) graph paper gives a sigmoidal curve which has a linear region from which good estimations of the sizes of unknown fragments can be made. Plotting the size in base pairs against 1/distance travelled has been reported to give an increased linear range for gels run at low voltage [16]. Gels can also be analysed by computer programs with data being inputted from a digitizing tablet which can increase the accuracy of the size determination. Fragments greater than about 20 kb, however, all move at the same rate when a constant electric field is applied and techniques to separate these by frequently re-orienting the electric field have been devised (see Chapters 11–13). In agarose gels of between 1 and 1.5% small fragments also travel with a constant mobility as their end-to-end length is less than the pore size of the gel [5].

The conformation of the DNA also affects the rate of travel. For double-stranded DNA plasmids, linear, nicked (open) circles and supercoiled DNAs move at different rates and the number of superhelical turns also affects supercoiled DNA mobility [17,18]. Electrophoresis of small plasmids such as the pUC series [19] in TAE buffer results in monomeric supercoils travelling furthest down the gel, with the linear and nicked (open) circular molecules running slower, the nicked circles being the slowest. The slower migrating forms often run amongst higher forms, dimers, trimers, etc; and it is often difficult to identify the different forms apart from monomeric supercoils. Linear molecules derived from a supercoiled plasmid can be identified by running a suitable digest of the plasmid in an adjacent track. Single-stranded DNA molecules normally run ahead of double-stranded counterparts. TBE buffer decreases the effective pore size of agarose and supercoiled plasmids migrate slower than in TAE buffered agarose [5].

The composition and ionic strength of the buffer and the applied field also affect the movement of the DNA. If the buffer is left out or if the ionic strength is too low then the DNA will not move into or through the gel; if the buffer is too concentrated then heat will be generated and gels have been known to melt! With regard to the applied electric field strength at low voltages the movement of linear DNA is proportional to the voltage but at high voltages the larger fragments move disproportionately quicker. Therefore large DNA fragments should be electrophoresed at 0.64 V/cm and smaller fragments at between 3.8 V/cm and 5 V/cm [2,5].

4. ELECTROPHORESIS BUFFERS

The most used buffers are based on tris-acetate (TAE) and tris-borate (TBE) (Table 2), the former having less buffering capacity and the latter causing problems with some subsequent DNA purification regimes. As TBE Buffer is also used for polyacrylamide gels, if both gel types are being used it may be more economical to stick to this one, but the author's preferred buffer is TAE. Buffers are made and stored as concentrated stock solutions.

5. POINTS FOR THE BEGINNER

5.1 Running a First Gel

The basic steps in preparing an agarose gel matrix involve dissolving agarose by heating in a suitable buffer solution and after cooling, pouring the still liquid solution into an appropriate mould or gel former.

Table 2.   Buffers for gel electrophoresis.

TAE electrophoresis buffer for native gels (Tris, Acetate, EDTA).

Made as a 50× stock and used at 1×.

50 × TAE	1× working solution
Required:	Molarities of components (1×)
242g Tris base	0.04 M Tris-acetate
57.1 mL glacial acetic acid	
100 mL 0.05 M EDTA (pH 8.0)	0.001 M EDTA
Make up to 1 L with deionised water	

TBE electrophoresis buffer for native gels (Tris, Borate, EDTA)
Make as a 10× stock and use at 1×. There are many variations on this recipe. Final concentration of 1× solutions can be as low as 29 mM Tris-borate with 2 mM EDTA.

10 × TBE	1× working solution
Required:	Molarities of components
43.2 g Tris base	0.09 M Tris-borate pH 8.3
27 g boric acid	
20 mL 0.5 M EDTA (pH 8.0)	1.25 mM EDTA
Make up to 1L with deionised water	

MESA (MOPS) electrophoresis buffer for denaturing formaldehyde gels
Make as a 10× stock and use at 1×. Some protocols use MESA at half the strength recommended here. It can now be bought ready prepared at 10× strength.

10× MESA (10× MOPS)	1× working solution
Required:	Molarities of components
82.4 g MOPS*	40 mM MOPS
8 g Sodium acetate (anhydrous)	10 mM sodium acetate
20 mL 0.5 M EDTA (pH 8.0)	1 mM EDTA

This can be made in DEPC treated water.
After autoclaving MOPS goes yellow but this normally does not affect the buffers properties.

---

*MOPS 3-(N-morphilino)propanesulphonic acid.

5.2   Don't Forget the Buffer!

The most common mistake by a beginner when running a first gel is to forget, or not to be told, to put the running buffer into the gel itself! The absence of buffer is normally immediately obvious when ,there is no current through a vertical gel and the DNA, and loading dye, fail to move into the gel. This absence of buffer is not always so obvious with a horizontal gel, especially if it has been left covered in electrophoresis buffer for some time before loading the samples. The DNA and dye may then move through the gel but the resulting DNA bands will be very poorly resolved and the gel will look a mess. Because electrophoretic separation has occurred in this situation it is not always obvious to the beginner that the poor separation that has occurred is due to the partial diffusion of buffer into the agarose during the standing period. As the buffer seldom saturates the gel poor resolution is the result.

5.3 Gel Equipment

Equipment for gel electrophoresis comes in a variety of shapes and sizes and is often manufactured in house. Horizontal gel tanks are currently the most common for agarose gel electrophoresis. These are made from transparent perspex, with two reservoirs for the buffer, and a shelf to support the gel. UV transparent perspex is used to make the gel former and this sits on the shelf. Gel tanks which allow a number of gels to be run one on top of the other are available (Hybaid). The ends of the gel former can be sealed with sticky tape, or special sealing plates/gaskets may be provided. Both in order to reduce costs and to save time the tank should be as small as possible while still allowing resolution of the DNA sample in as short a time as possible. With regard to the size of the gel the material being analysed must obviously be considered, simple plasmid digests with only 2–3 bands can be resolved in 30 min in small minigels. Separation of far more complex DNA, e.g. digests of human genomic DNA may require over 24 h for sharp band definition when subsequently probed. PFGE gels (Chapters 11–13) may require 48 h or longer. The gel tank should allow for the recycling of buffers if this is thought necessary during long runs, particularly with TAE buffer, and for safety reasons it should not be possible to remove the lid while the power supply is connected.

Vertical gel apparatus is more common for separations involving polyacrylamide which requires air to be excluded during the polymerisation process. If vertical apparatus is used for agarose gel electrophoresis problems arise with dilute (<0.8%) gels which require a roughened back plate to allow the agarose to adhere, and plugs of

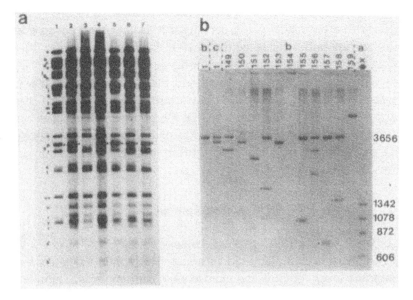

Figure 1. (a) Autoradiograph of radiolabelled Herpes simplex virus DNA digested with BamHI and seperated through a 1.0% agarose gel. Fragment a is 12474 bp, p is 3579 and z is 1840 bp. (b) 1% agarose gel of digested pAT153 recombinant clones [31].

acrylamide or porous solid membranes are required to support the bottom of the agarose gel between the glass plates. This support is particularly important when the gel is being removed from the apparatus after electrophoresis. Vertical gel apparatus, especially for use with polyacrylamide gels, has improved considerably in recent years particularly with regard to preventing the gel matrix from leaking out before it can polymerise

5.4   Preparation of an Agarose Gel

The following procedure applies to any agarose gel. Examples of agarose DNA gels can be seen in Figures 1 and 2.

1. Weigh out the appropriate amount of agarose according to the volume and concentration of the required gel. The concentrations of gels for the separation of DNA fragments of different sizes are shown in Table 1.

2. Add the agarose to the appropriate volume of 1 × TAE (Table 2) gel buffer in a conical flask. Mark the level of the liquid if the agarose concentration is critical so that evaporated water can be replaced. The top can be blocked with non-absorbent cotton wool or a paper tissue.

3. Dissolve the agarose by bringing the buffer to the boil and simmering. Take care, higher (2%+) concentrations of agarose can caramelise if overheated. If a microwave is being used it is safer to do this on the simmer setting rather than high, which may superheat the vessel and agarose causing the agarose to shoot out from the container if it is shaken slightly. A face mask and gloves should be worn. The state of the agarose can be checked by holding it up to the light and looking for undissolved globules.

4. If the agarose percentage is critical replace any evaporated water by reference to the mark made on the vessel. For most purposes this is not critical if you do the same routine every time and many gels are slightly more concentrated than people state.

5. Cool the agarose to 55°C.
   N.B. If very concentrated agarose solutions are being poured (e.g. 4%) the agarose must be poured almost immediately it has boiled as it sets quickly and a non-homogeneous gel may result. If vertical agarose gels are being run it is sensible to warm the glass plates prior to pouring in hot concentrated agarose in order to prevent the plates cracking.

6. Pour the cooled agarose into a gel former with ends sealed and add one or more combs. Leave to set.

7. Remove the sealing tape. Some people remove the comb at this stage.

8. If the gel has been poured into a former that is in the gel tank add 1× TAE buffer to just below the top of the agarose, or to just over the top. This is personal preference but for some gels it may be important to keep the top dry when loading samples. I usually remove the comb once buffer just covers the top of the gel so that the wells fill with buffer as the comb is removed.

5.5   Loading and Running the Gel

Table 3.   Sample loading buffers.

(i) Loading buffers for native gels
Six times strength loading buffers are normally made up in the appropriate electrophoresis buffer or water and stored at 4°. Ficoll containing buffers are recommended.

6 × Buffer a
0.025% Bromophenol blue
0.025% xylene cyanol FF (optional)
15% Ficoll 400

6 × Buffer b
0.025% Bromophenol blue
0.025% xylene cyanol FF (optional)
30% glycerol

(ii) Loading buffers for formaldehyde denaturing gels
Make up as 10× concentrates

10 × Buffer c
0.4% Bromophenol blue
0.4% xylene cyanol FF (optional)
50% glycerol
1 mM EDTA

10 × Buffer d
0.4% Bromophenol blue
0.4% xylene cyanol FF (optional)
30% Ficoll 400
1 mM EDTA

---

1. Prepare the DNA samples using a loading buffer from Table 3. The use of Ficoll is recommended as it minimizes trailing and spread of the DNA into adjacent tracks. Dilute the appropriate sample loading buffer with DNA and water. If you are running your first gel it is useful to practice loading buffer into one or two spare wells. This allows you to determine how much sample can be loaded into each well before you use valuable material. The amount of sample that can be loaded varies according to the number of teeth on the comb used and the thickness of the gel. Some small minigels can take only 10 μL per well whereas large thick gels may take 100 μL or considerably more if only one large well is being used for preparative work. 25 μL to 40 μL are routinely loaded for many

purposes. The amount of DNA loaded depends on the number of bands to be seen and the purpose of the gel. 100 ng of λ DNA (50 kb) digested with HindIII, which is often used as a marker digest and produces eight fragments, should enable the six larger bands to be easily seen. To see the small 560 bp fragment, however, 400 ng may be required, depending on how long the gel is stained and the strength of the UV system. With modern probe generating protocols 2 μg of mammalian DNA can be used to detect single copy genes but many would load 5 to 10 μg.

2. Load the samples into the wells using a micropipette and appropriate tip.

    The buffer should sink into the well when the tip is placed just below the surface but some people place the tip to the bottom of the well and withdraw it as the sample comes out. Take care not to allow sample to overflow into the next well. For most purposes the same tip can be reused after washing two or three times with buffer but IF THE GEL IS TO BE BLOTTED OR THE DNA PURIFIED FOR LATER USE, e.g. CLONING, ON NO ACCOUNT USE THE SAME TIP TWICE.

3. Place the lid on the gel apparatus, check that the electrical connections provide the correct polarity of current flow. The DNA is negatively charged and will travel towards the positively charged anode. Electrophorese the gel at up to 5 V/cm depending on the fragment sizes you wish to separate.

4. When the bromophenol blue has travelled to within about 1 cm from the end of the gel stop the electrophoresis by switching off the power and then transfer the gel to ethidium bromide for staining (see later). DNA separated on a 1% gel can be seen in Figure 1.

5.6 Very Dilute Agarose Gels

When pouring very dilute agarose gels these can be made more manageable by pouring a concentrated base of higher concentration, e.g. 2% agarose, and then pouring the more dilute gel on top once the base has set.

6.   EXAMPLES OF GEL SYSTEMS

6.1   High Concentration Composite Agarose Gels

One percent agarose gels provide very good resolution for a wide range of DNA molecules. More concentrated agarose gels can be used to separate small DNA fragments in preference to polyacrylamide gels. In particular, composite agarose gels consisting of mixtures of agaroses with high resolution but low gel strength and low resolution but higher strength can provide high resolution of small fragments that would traditionally have been separated by polyacrylamide gels [7,2 0,2 1]. Highly purified Nusieve and Seakem agaroses (FMC) used at a 3 : 1 ratio in 2% to 4% gels can provide good resolution of small DNA fragments such as the products of PCR reactions. The Nusieve agarose provides high resolution, but is a low gelling temperature agarose giving low gel strength. The Seakem agarose has lower resolution but gels at higher temperature and strengthens the

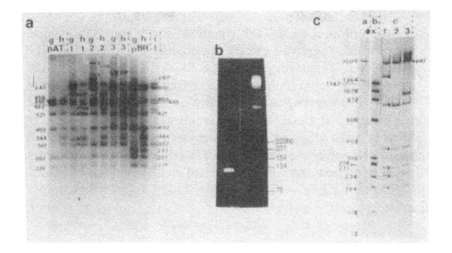

Figure 2.   (a) 4% agarose gel showing separation of digested DNAs. (b) 4% composite agarose gel showing the reolution of a 123 bp PCR product. The marker is a 1 kb ladder (Life Sciences Ltd) [30]. (c) 4% polyacrylamide gel showing the resolution of digested plasmid and phage $\phi \times 174$ DNAs.

agarose matrix. These gels, however, can only be used for analytical work, higher grade agaroses such as Nusieve GTG (FMC) being available for preparative work. The gels are prepared exactly as described above, mixing the appropriate amounts of the two agaroses prior to dissolving in buffer (they can be bought ready mixed, FMC). 4% gels can easily resolve fragments down to around 20 bp, the size of many PCR primers, with the range of separation reported to be 10 to 1000 bp. Figure 2 shows the resolution obtained by 4% agarose and polyacrylamide gels.

When loading PCR products onto horizontal agarose gels it is important to keep buffer off the top of the gel until the samples have all been loaded and electrophoresed in to the gel for a short distance. The samples should not be allowed to overflow from one well to another. When the gel is placed into the gel tank the buffer should be added until it is near the top of the agarose, the sample loaded, electrophoresed into the gel for about 1 cm and only then should the buffer be added to cover the agarose. It has been reported that overlaying the gel with too much buffer (>1-mm depth) may allow small fragments to migrate from the gel into the buffer [6].

6.2 Separation of DNA Fragments in Non-denaturing Polyacrylamide Gels

Although agaroses are available which can replace polyacrylamide there may be situations in which separation of DNA through polyacrylamide is preferred. A variety of gel sizes are available for running polyacrylamide gels, however the main ones commercially available (apart from those for sequencing gels) produce gels which are around 16 cm wide and 16 or 20 cm long with minigel systems producing gels around 10 cm wide by 7 cm long. Pharmacia produces even smaller gels for its Phastsystem. It is important if home made or older apparatus is being used that the glass plates and spacers allow the joints to be tightly sealed, particularly at the corners, as acrylamide solutions are very good at leaking through the smallest gap. Many commercial kits are well designed with good seals, making it unnecessary to use sticky tape. Spacers for polyacrylamide gels vary in thickness from 0.5 to 2.0 mm, thicker gels presenting problems with heat generation during electrophoresis which can lead to smiling of the DNA bands.

Table 4. Reagents for polyacrylamide gels.

30% acrylamide:	29 g acrylamide
	1 g N,N´-methylene
	   bisacrylamide
	Deionised water to 100 mL
	Filter through Whatman number
	   1 filter paper.

10% Ammonium persulphate:	1 g in 10 mL of deionised
	   water.
	Can be stored for up to 1 week
	   at 4°C .
	Some prefer to make this fresh
	   each time.

TBE, See Table 2	
Loading buffer, see Table 3.	

30% acrylamide can be prepared in house (Table 4) but can also now be bought as a pre-mixed solution with various ratios of acrylamide to bis-acrylamide from many reagent suppliers. This reduces the hazards associated with handling the powder, unpolymerised acrylamide being a cumulative nerve toxin. The powder should only be weighed and handled using gloves, face mask and in a suitably ventilated hood or glove box designed for this type of material. Because of the possibility that even polymerised gels may contain some unpolymerised material, gloves should always be worn at this stage as well.

6.2.1   Preparation of the glass plates

The glass plates used for PAGE should be clean and grease free. Successive washing in acid (0.1% hydrochloric acid) and alkali (5% potassium hydroxide) followed by thorough rinsing in deionised water can be performed. Some would recommend a detergent wash after the alkali wash, prior to rinsing well, first in tap water and then deionised water, others only do an ethanol wash and to a large extent the amount of washing depends on the state of the plates! Gloves should be worn to keep fingerprints off the glass and the plates dried on fluff free towelling. (Many people manage to run perfectly acceptable polyacrylamide gels without this thorough preparation!)

In order to be able to remove the gel after electrophoresis one, or both glass plates can be siliconised. Again this is not essential and one should judge the need from personal experience. The glass plates and the spacers must be arranged together and sealed with either 3M yellow electrical tape (Life Sciences: 1032ST) or other methods, such as bulldog clips as recommended by the manufacturer. Particular attention must be paid to the bottom corners of the plates to make sure any tape is well pressed down. Running a fingernail along overlaps can help to seal these joints which are especially prone to leaking. Some people seal the plates by running a little 1% agarose in 1× TBE down the side and along the bottom of each gel but it can be difficult to get this to run into very narrow gels.

6.2.2. Preparation of the polyacrylamide

A conical flask can be used to mix the reagents for the required concentration of the polyacrylamide gel according to the quantities stated in Table 5 [2,6]. For 16 cm by 16 cm gels with 0.2-mm spacers about 70 mL of acrylamide should suffice. TEMED (35 μL/100 mL) should be added last (and should be dispensed in a fume hood), the reagents carefully mixed and poured down the larger back plate into the gel cavity. A 50 mL syringe without a needle can be used to load the acrylamide between the glass plates. If the gel is not in a vertical holding apparatus lay it down at a slight angle to reduce the possibility of leakage. Immediately place the comb in between the plates making sure air bubbles are excluded, clamp the comb in place if necessary and leave the gel to set (30 to 60 min). Slight loss of acrylamide can be compensated for by adding acrylamide solution to the top of the gel, down the side of the comb. When the gel has polymerized you can see a wavy (Schlieren) pattern beneath and around the teeth of the comb.

Table 5. Components for polyacrylamide gels.

Components[a]	Final acrylamide concentration					
	3.5%	4%	5%	8%	12%	20%
30% Acrylamide	11.6	13.3	16.6	26.6	40	66.6
Water	77.7	76.0	72.7	62.7	49.3	22.7
103 TBE	10	10	10	10	10	10
10% Amm pers.	0.7	0.7	0.7	0.7	0.7	0.7
TEMED[ab]	0.035	0.035	0.035	0.035	0.035	0.035

[a]Components are in mL for 100 mL of solution
[b]Can be used diluted 10× instead of neat and volumes adjusted accordingly

6.2.3   Failure of polymerization

If the gel does not polymerize it could be that there was too much oxygen in the acrylamide and degassing of the solutions may be required. This must be done in a way that prevents exposure to aerosols of acrylamide. If the area around the wells does not set then it may be that the comb was not providing a good seal. A common cause of failure to polymerize properly is that the ammonium persulphate solution is too old or the stock has gone off and needs replacing.

If not required immediately, the gel can be stored at 4°C for a few days.

The sealing tape and clips can be removed at this time, particularly the sealing mechanism from the bottom of the gel, and the gel placed into the gel tank. Do not remove the tape or clips if the gel is to be stored. Fill the tanks with buffer, check for leaks and remove the comb. Immediately flush out the wells using a syringe with a needle, or suitable tubing in order to remove unpolymerised acrylamide which can polymerise and block or distort the wells, causing problems when loading the samples. Some would remove the comb before placing the gel in the tank.

Remove any air bubbles from between the glass plates at the bottom of the gel using a bent Pasteur pipette, or syringe with attached tubing (the latter is safer!).

The DNA samples are loaded as for the agarose gels. A suitable loading buffer is described in Table 3. A micropipette tip may suffice but the tip of a Hamilton syringe is easier to insert between narrowly spaced plates. The volume loaded varies according to the gel and tooth size but 20 to 30 µL is usually easily loaded. Electrophoresis is normally carried out at 8 V/cm; too high a voltage may cause overheating and even denaturation of the DNA.

After electrophoresis switch off the power pack, disconnect the gel and pour away the buffer. Remove any tape by using a scalpel blade, carefully remove the notched or shorter plate by levering it up with a spatula (check that the gel is attached to the other plate, if not turn the gel over and remove the other plate).

Place the gel into ETBR for staining. For ease of manipulation the gel can be left on the glass plate. After staining rinse well and visualise with UV light. Figure 2c shows a 4% polyacrylamide gel.

6.3 Separation of RNA on Denaturing Agarose Gels

Analysis of RNA normally takes place using denaturing gels which use glyoxal/DMSO, methylmercuric hydroxide or formamide/formaldehyde as denaturants. The use of formaldehyde provides good resolution for many purposes and this is the author's preferred system, although glyoxal/DMSO gels may give sharper bands (2). The use of methyl mercuric hydroxide should be avoided on safety grounds unless other systems do not provide adequate denaturation (e.g. GC rich sequences).

Prior to running the RNA sample on a formaldehyde gel a quick method to assess the quality of a cellular RNA preparation is to run a small aliquot on a freshly made 1.5 to 2% agarose minigel at pH 7.5. By running the dye front only one third of the way down the gel

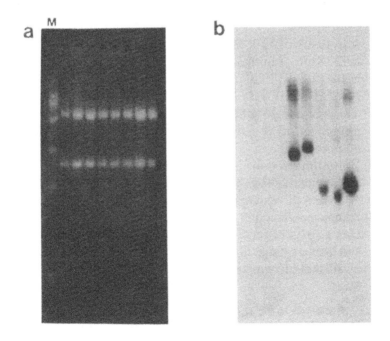

Figure 3. (a) 1.5% agarose–formaldehyde gel of total cellular RNA stained with ETBR. RNA markers are in the track marked M and the 28S and 18S ribosomal RNA bands from COS cells can be seen in all other tracks. (b) A Northern blot of a similar gel to that in (a). The probe has detected mRNAs between approximately 2500 and 900 nucleotides in size.

followed by ETBR staining and UV visualisation, the ribosomal bands should be clearly visible without any trace of smearing. Care must be taken to prevent RNAse contamination in this and all RNA work and details of how to deal with RNAses are discussed in Chapter 9. The protocol described is based on that in Ref. [2].

6.3.1   Preparation of the gel

The mobility of DNA in formaldehyde gels has been analysed [22] and 1% to 1.5% gels provide good resolution of many RNA species of interest. Figure 3c shows an autoradiograph of a Northern blot of RNAs transferred from a 1.5% gel. The gel buffer is made from 1 × MESA (Table 2) and the gel contains 2.2 M formaldehyde.

(a) For 100 mL of 1.5% agarose:
Dissolve 1 g of agarose in 72 mL of water
Cool to 60°C
Add 10 mL of 10 × MESA buffer
Add 18 mL of formaldehyde
N.B. THE STEPS INVOLVING FORMALDEHYDE MUST BE CARRIED OUT IN A FUME HOOD.
The agarose-formaldehyde buffer solution should then be poured into the gel former, the comb inserted and the gel allowed to set.
(b) Cover the gel with 1 × running buffer when set.
Subsequently the gels should always be covered with buffer and run in a well ventilated area.
(c) Sample preparation
Up to 30 μg of total cellular RNA can be loaded into one well and it may be that the RNA preparation will need concentrating by ethanol precipitation to keep the volume small.
Mix:

RNA (up to 20 μg to 30 μg)	4.5 μL
10× MESA	2.0 μL
Formaldehyde	3.5 μL
Formamide	10.0 μL
Total	20.0 μL

Heat the sample for 15 min at 65°C, place on ice and when cool spin briefly to concentrate all of the solution at the bottom of the tube. Add 1 μL of 10 × RNA loading buffer

(Table 3) prior to loading the samples in the wells. RNA markers should be treated in the same way.

Mixtures of RNAs for use as size markers for RNA gels can be bought from suppliers (e.g. Life Sciences Inc.) and these are normally placed in the outside wells. Alternatively if the sample is total cellular RNA the positions of the ribosomal bands can be used as markers. HeLa cell ribosomal RNAs are 6333 (28S) and 2366 (18S) nucleotides in length.

A most efficient way of visualising the RNA bands after electrophoresis is to add 1 µg of ETBR (2 µL of a 0.5 mg/mL solution) to each sample prior to loading the gel [23]. This slightly reduces the efficiency of Northern transfer but removes the need to cut off and stain the marker tracks separately to any samples which are to be blotted. If problems are encountered with Northern transfers when ETBR is added to the sample omit it and stain bands which are not to be transferred after electrophoresis.

6.3.2   Running the gel

The gel can be prerun prior to loading samples at 5 V/cm for 5–15 min. This equilibrates the buffers within and outside the gel and heats up the gel to its working temperature [2,5]. In addition the buffer from the tanks can be mixed at the end of 1–2 h of electrophoresis. However we find that perfectly good gels can be obtained without these steps.

When the samples have been loaded the gel is run at 3–4 V/cm until the dye has travelled about three quarters of the way through the gel.

If the gel has been prestained with ETBR the RNA can be visualised using UV light and photographed, or marker tracks removed, stained separately to the bulk of the gel and photographed. The most common procedure to follow the electrophoretic separation of RNA is to blot the RNA onto nylon or nitrocellulose (Northern blot) and probe for specific species of RNA. If repeated probings are to be performed the RNA should be blotted onto positively charged nylon membrane according to the protocol of the membrane manufacturer [24]. Prior to blotting the gel should be rinsed in water to remove the formaldehyde, some recommend treatment of the water with DEPC (but take care: this is a suspected carcinogen!). An ETBR stained RNA gel can be seen in Figure 3a and a Northern blot of the same gel in figure 3b.

7.   STAINING AND VISUALIZATION OF DNA AND RNA
IN GELS

7.1   Use of Ethidium Bromide

The most common way of visualising DNA and RNA in agarose gels is to use the fluorescent dye ethidium bromide (ETBR) which intercalates between the bases in both DNA and RNA molecules. The intercalated ETBR absorbs UV radiation at 300 and 360 nm, and transmits it at 590 nm causing a red-orange fluorescence.

ETBR is a powerful mutagen and should be handled with care. It can now be bought in aqueous solution at 500 µg per mL or as 100 mg tablets (SIGMA and others) reducing potential handling problems. The safest way to use ETBR and keep contamination to a minimum is to run the gel and then to stain for 30 min using 3 to 4 gel volumes of water containing 0.5 µg/mL of ETBR in containers reserved for staining. Before examination of the stained gel is undertaken it should be washed in clean water or buffer. If the bands appear faint after a short staining the gel can be left for a longer period (e.g. overnight).

The author has found that it is rarely necessary to destain gels in water to reduce background fluorescence. However washing the gel and leaving it in water for a time does reduce the level of ETBR in the gel which reduces the hazard from this mutagen and destaining should be performed on safety grounds. Methods for disposing of ETBR are described later.

If RNA is still present in DNA gels and is obscuring bands in an important region of the gel, RNase can be added. As the RNA is degraded the digestion products diffuse out of the gel clearing the area.

Plastic scoops used for removing unwanted material your pets may leave behind are very useful for moving small gels on and off transilluminators. We try to keep ETBR off gel apparatus, including the trays supporting the gel, as much as possible so we remove the gels from these before staining. UV protective face masks (not any old face mask; check that it does not transmit UV!) and gloves should always be worn and the DNA exposed to UV irradiation, particularly short wave radiation, for as little time as possible to prevent damage. UV transilluminators can be bought which emit UV light at wavelengths of 254, 302 and 366 nm. 254-nm sources should be avoided as these can damage the DNA excessively and the most useful wavelength as it produces less damage but gives strong fluorescence is 302 nm. In agarose gels 1 ng of DNA can be visualized by using ETBR but for polyacrylamide gels around 10 ng is the minimum due to quenching of fluorescence.

7.2   Staining with Brilliant Cresyl Blue

A method of staining DNA in agarose gels using the dye brilliant cresyl blue has recently been described [26]. It is reported that 0.025 µg of DNA can be detected, and the production of permanent records does not require a UV transilluminator. This method may be of value for laboratories where a UV transilluminator and camera are not available. This protocol is best performed with smaller gels (e.g. 7 × 9 cm or less) as a centrifugation step is involved. The method is taken from Santillan-Torres and Ponce-Noyola, 1993 [26].

After electrophoresis the gel is immersed in a staining solution containing 0.04% brilliant cresyl blue (SIGMA) in 20% ethanol for 15 to 60 min, depending on the concentration of DNA. The gel is then washed with distilled water for 10 s to remove excess dye, washed with 70% ethanol for 5 s and twice with distilled water. A permanent record can be made by photography, or a copy made by transferring the dye to 180–200 g coated art paper ('couche' paper). To transfer the dye from the gel it is first placed on a wet cellophane sheet lying on a stack of paper towels. Three pieces of wet filter paper and a glass plate are placed on top and the sandwich centrifuged at 2000 r.p.m. for 15 min at room temperature, using a rack in a Beckman TJ-6 centrifuge rotor. After centrifugation the gel is removed from the cellophane by immersion in water for 5 to 10 s. The gel is next placed on a sheet of white coated art paper, and making sure that there are no air bubbles, it is recentrifuged as above.

After centrifugation the dye should have entered the paper to produce a permanent record of the gel. Carefully remove the gel and dry the paper at room temperature. However, if some dye remains in the gel, do not remove it, wrap the gel and paper in polythene or nylon, and apply pressure by using a weight for 15 min. Carefully remove the gel by soaking in water for 10 s and dry the paper with the transferred dye at room temperature. The gel can be restained and up to two further copies made.

DNA bands in polyacrylamide gels can also be stained using methylene blue [5].

8. PHOTOGRAPHIC RECORDS OF GELS

DNA and RNA visualized by UV irradiation can be recorded on photographic film by use of a Wratten 22A or 23A filter over the camera lens to filter out the red light from the lamp, and a suitable black and white film. Polaroid camera based systems are popular, using either Polaroid Type 665 (positive/negative) or 667 (positive) film to quickly provide a permanent record, although video cameras with CCD lenses are now starting to be used (UVP). These are expensive to buy but the image can be stored on computer disk and copies cost only a few pence to produce. A ruler can be placed along the side of the gel prior to photography so that the distances the bands have migrated can be directly recorded.

9. DISPOSAL OF ETHIDIUM BROMIDE

The ETBR used to visualise DNA and RNA is a strong mutagen and should only be handled with gloves and weighed out using an appropriate glove box or hood. Disposal of ETBR should be performed by a route which destroys its mutagenic activity without creating additional hazards. A number of protocols have been devised to dispose of ETBR [27,28,29] (compiled in [2]), and one for dilute [29] and one for concentrated solutions [27] are described below.

9.1   For Concentrated Solutions

After [29].
1. Dilute the solution containing ETBR to less than 0.5 mg/mL.
2. Add 0.2 vol of freshly diluted 5% hypophosphorous acid (CARE, THIS IS CORROSIVE!) and 0.12 vol of fresh 0.5 M sodium nitrite (69 g/L, w/v)).
3. Mix well and carefully; the pH must be <3.0.
4. After 24 h add a large excess of 1 M sodium bicarbonate and discard.

9.2 For Dilute Solutions

This protocol can be used for dilute solutions containing less than 0.5 µg/mL ETBR such as gel buffers [27].

1. Add 100 mg of powdered activated charcoal to each 100 mL of solution.
2. Store, with occasional shaking.
3. Filter through Whatman number 1 filter paper, discard the filtrate.
4. Seal filter and charcoal and dispose for incineration as hazardous waste.

ACKNOWLEDGMENTS

Thanks to E. Norman for critically reviewing the manuscript.

REFERENCES

[1] Rickwood, D., and Hames B.D. (eds), *Gel Electrophoresis of Nucleic Acids: A Practical Approach*. IRL Press, Oxford, 1982.

[2] Sambrook, J., Fritsch, E. F., Maniatis, T. (eds), *Molecular Cloning: A Laboratory Manual*. Cold Spring Harbor Laboratory Press, Cold Spring Harbor, N.Y. 1989.

[3] Berger, S.L. and Kimmel, A.R. (eds), *Guide to Molecular Cloning Techniques* (Methods in Enzymology, Vol. 152). Academic Press, New York, 1987.

[4] Mosher, R. A., Saville, D. A. and Thorman, W. (eds), *The Dynamics of Electrophoresis*. VCH, Weinheim, 1992.

[5] Noolandi, J. Theory of DNA gel electrophoresis. In Chram-buch, A., Dunn, M. J. and Radola, B. J. (eds), *Advances in Electrophoresis*, Vol. 5. VCH, Weinheim, 1992, pp. 1–57.

[6] Ogden, R. C. and Adams, D. A. Electrophoresis in agarose and acrylamide gels. In Berger, S. L. and Kimmel, A. R. (eds), *Guide to Molecular Cloning Techniques* (Methods in Enzymology, Vol. 152), Academic Press, New York, 1987, pp. 61–87.

[7] Resolutions. FMC Newsletter, 6 (1990) No. 2.

[8] Stellwagon, N. C. Anomalous electrophoresis of deoxyribo-nucleic acid restriction fragments on polyacrylamide gels. Biochemistry 22 (1983) 6186–6193.

[9] Stellwagon, N. C. Electrophoresis of DNA in agarose and polyacrylamide gels. Adv. Electrophoresis, 1 (1987) 177–228.

[10] Aaij, C. and Borst, P. The gel electrophoresis of DNA. Biochim. Biophys. Acta, 269 (1972) 192–200.

[11] Fisher, F. P. and Dingman, C. W. Role of molecular conformation in determining the electrophoretic properties of polynucleotides in agarose-acrylamidce composite gels. Biochemistry, 10 (1971) 1895–1899.

[12] Lerman, L. S. and Frisch, H. L. Why does the electrophoretic mobility of DNA in gels vary with the length of the molecule. Biopolymers, 21 (1982) 995–997.

[13] Ogston, A. G. The spaces in a uniform random suspension of fibres. Trans. Faraday Soc., 54 (1958) 1754–1757.

[14] Stellwagon, N. C. Accurate molecular weight determination of deoxyribonucleic acid restriction fragments in agarose gels. Biochemistry 22, (1983) 6180–6185.

[15] Helling, R.B., Goodman, H. M. and Boyer, H.W. Analysis of R.EcoRI fragments of DNA from lambdoid bacteriophages and other viruses by agarose-gel electrophoresis. J. Virol., 14 (1974) 1235–1244.

[16] Southern, E. M. Detection of specific sequences among DNA fragments separated by gel electrophoresis. J. Mol. Biol., 98 (1975) 503–517.

[17] Thorne, H.V. Electrophoretic characterization and fractionation of polyoma virus DNA. J. Mol. Biol., 24 (1967) 203–211.

[18] Johnson, P. H. and Grossman, L. I. Electrophoresis of DNA in agarose gels. Optimizing separations of conformational isomers of double- and single-stranded DNAs. Biochemistry 16, (1977) 4217–4225.

[19] Vieira, J. and Messing, J. The pUC plasmids: an m13mp7-derived system for insertion mutagenesis and sequencing with synthetic universal primers. Gene, 19 (1982) 259–268.

[20] Asaki, R., Chung, D.W., Ratnoff, O.D. and Davie, E.W. Factor XI (plasma throboplastin antecedent) deficiency in Ashkenazi Jews is a bleeding disorder that can result from 3 types of point mutations. Proc. Natl. Acad. Sci. USA, 86 (1989) 7667–7671.

[21] Green, E.P., Tizard, M.L.V., Moss, M.T., Thompson, J., Winterbourne, D.J., McFadden, J. J. and Herman-Taylor, J. Sequence and characterisation of IS900, an insertion element identified in a human Chrohn's disease isolate of Mycobacterium paratuberculosis. Nucleic Acids Res., 17 (1989) 9063–9073.

[22] Lehrach, H., Diamond, D., Wozney, J. M. and Boedtker, H. RNA molecular weight determinations by gel electrophoresis under denaturing conditions; a critical evaluation. Biochemistry 16, (1977) 4743–4751.

[23] Kroczek, R.A., and Siebert, E. Optimisation of northern analysis by vacuum blotting, RNA-transfer visualisation and ultra violet fixation. Anal. Biochem., 184 (1990) 90–95.

[24] Reed, K.C. Nucleic acid hybridizations with positive charged-modified nylon membrane. In Dale, J.W. and Sanders, P.G. (eds), *Methods in Gene Technology*, Vol. 1, JAI Press, London, 1991, pp. 127–160.

[25] Sharp, P. A., Sugden, B. and Sambrook, J. Detection of two restriction endonuclease activities in Haemophilus parainfluenzae using analytical agarose-ethidium bromide electrophoresis. Biochemistry, 12 (1973) 3055–3063.

[26] Santillan-Torres, J. L. and Ponce-Noyola, P. A novel stain for DNA in agarose gels. Trends Genet., 9 (1993) 40.

[27] Lunn, G. and Sansone, E. B. Ethidium bromide: Destruction and decontamination of solutions. Anal. Biochem., 162 (1987) 453.

[28] Quillardet, P. and Hofnung, M. Ethidium bromide and safety-readers suggest alternative solutions. Trends Genet., 4 (1988) 89.

[29] Bensaude, O. Ethidium bromide and safety-readers suggest alternative solutions. Trends Genet., 4 (1988) 89.

[30]   Green, J. The role of semen in the transmission of genital human papillo-
       maviruses. Ph.D Thesis (1993) University of Surrey, 1993.
[31]   Sanders, P.G. Analysis of mutations produced at random and in specific
       regions of the HSV genome. Ph.D Thesis, University of Glasgow, 1993.

Chapter 3

PREPARATIVE GEL ELECTROPHORESIS

John C. Maule

OUTLINE

Methods in Gene Technology, Volume 2, pages 43–63
Copyright © JAI Press Ltd
All rights of reproduction in any form reserved.
ISBN: 55938-264-3

1.   INTRODUCTION

Agarose gel electrophoresis has proved to be one of the most important and widely used methods for the separation of DNA molecules. As well as a powerful analytical technique, it is used extensively as a tool for the preparation of DNA fragments for genetic manipulation. DNA recovered from agarose gels can be used in cloning procedures and in the generation of labelled probes. Any technique for the preparation of DNA has to satisfy several criteria and the plethora of methods designed to accomplish the recovery of DNA from agarose gels is evidence that no single approach is entirely satisfactory. The following criteria can be used to judge the efficacy of the various methods for preparative gel electrophoresis.

(1) The most successful approaches must produce DNA of sufficiently high purity to take part in further enzymatic manipulations. DNA should be pure enough to act as a substrate for cleavage by restriction endonucleases and subsequent ligation to produce molecules capable of transformation. The generation of probes for hybridisation demands that DNA should be of sufficient quality to be efficiently labelled by nick translation, random priming or end labelling. In general, the purity of DNA recovered from gels is influenced by the quality of the agarose used in their preparation. Several high grade agaroses are now commercially available and the

extra cost involved in their purchase is generally rewarded by the recovery of DNA requiring no further purification. Several ultra pure low melting temperature agaroses allow the use of 'in-gel' enzymatic manipulations [1,2], which avoid the separation of DNA from agarose.

(2) A good yield of DNA following recovery from agarose gels is an essential prerequisite for any successful preparative method. The yield is influenced by the concentration of DNA in the starting material as well as the size of the molecules. In general, most methods do not recover very short molecules (<500 bp) and yields of very large fragments are usually reduced. Any further steps necessary to increase the purity of the recovered DNA will almost certainly lead to a reduction in the final yield.

(3) Methods that require few manipulations or are intrinsically rapid offer considerable advantages, especially if they result in good yields of high purity DNA. Some approaches, although lengthy, require a small amount of operator time and these methods are particularly suited to the processing of multiple samples.

(4) Consistent reproducibility is an important attribute of any successful purification procedure. The refinement of several published methods and the use of high quality reagents have increased the reliability of several of the techniques. Doubtless the availability of high quality agarose, with minimum batch to batch variation, has contributed significantly to the success of many approaches.

In this chapter I aim to present several methods which fulfil these criteria. These methods rely on different principles and some of them have not yet found their way into the standard laboratory protocol manuals [3]. The reader may find their inclusion a refreshing alternative to the frequently described approaches, which can be found described in the literature [4].

2. PREPARATIVE AGAROSE GELS

2.1 Principle

Optimising the fractionation of DNA, with the intention of recovering material from an agarose gel, depends on a number of parameters, including loading the correct concentration of DNA and choosing the most appropriate agarose and running conditions. The detection and excision of the fragment for recovery must be performed correctly to minimise contamination or damage to the DNA. The analytical use of agarose gels to separate DNA is covered elsewhere in this book (Chapter 2) and readers, new to the subject, should first consult that chapter.

2.2 Materials

2.2.1   Gel running buffers

1 × TAE
40 mM Tris
20 mM sodium acetate
1 mM EDTA
Make up as a 20 × stock, adjusting the pH to 8.0 with acetic acid.

0.5 × TBE
45 mM Tris
45 mM boric acid
1 mM EDTA
Make up as 5 × stock. There is no need to adjust the pH, which should be about 8.

2.2.2   Ethidium bromide.

Make up a stock at 10 mg/mL and store in the dark by wrapping aluminium foil around the container.
CAUTION: Ethidium bromide is mutagenic and should be treated with extreme care. Avoid any contact with the skin or accidental inhalation. Biorad make ethidium bromide tablets which avoid the need to weigh out the powder. Each tablet makes 11 mL of stock solution. Methods for the safe disposal of ethidium bromide are available [3].

2.2.3   Agarose.

Several types of agarose are suitable for preparative work (see Section 2.4 for an explanation of agarose specifications).
(a) Ultra pure standard melting temperature agaroses are characterised by a high gel strength ($>1000 \, g/cm^2$ for a 1% gel), low

sulphate content (usually <0.15%) and, most importantly, no detectable DNase or RNase activity. DNA molecules above about 0.2 kb are efficiently separated, the range of separation being determined by the agarose concentration and to some extent by the buffer. FMC Seakem GTG and Boehringer agarose MP are two examples, the latter being of a very high specification and having the additional advantage of high gel strength (>1800 g/cm^2 for a 1% gel) so that it can be used at low concentrations. This is a particularly attractive feature for pulsed gel electrophoresis, where low agarose concentrations result in significantly reduced run times. These agaroses are economical and should be chosen for separating DNA >0.2 kb in size, when there is not a requirement to recover the DNA by melting the gel.

(b) Ultra pure low melting temperature (LMT) agaroses melt at about 65°C and should be used in applications where DNA is recovered from molten agarose or for 'in gel' enzymatic applications. These agaroses have the same low sulphate and EEO specifications as ultra pure standard melting temperature agarose but are characterized by a low gel strength (usually about 200 g/cm^2 for a 1% gel), making them rather fragile to handle. These agaroses are expensive but can be used economically by using them only in parts of the gel where DNA samples are to be recovered. Thus a gel can be cast using standard melting temperature agarose and then the section destined to contain the DNA samples to be recovered, excised and re-filled with LMT agarose. This approach has the additional advantage that the more rigid gel margin improves the handling characteristics of the gel. These agaroses are capable of separating DNA molecules above about 0.1 kb in size. FMC SeaPlaque GTG and BRL Ultrapure low melting temperature agarose are two recommended products.

(c) FMC NuSieve GTG is an agarose designed to separate DNA in the 10–1000 bp size range, obviating the need to use polyacrylamide. It is normally used at a concentration between 3–4% and has the same sulphate and EEO characteristics as the other ultra pure agaroses. NuSieve GTG is a LMT agarose and so can be used for 'in gel' nucleic acid manipulations or applications requiring molten agarose. 4% NuSieve has a gel strength of >500 g/cm^2, so these gels are rather brittle and should be handled carefully. FMC recommend a 3 : 1 mix of NuSieve GTG and Seakem GTG, has a separation range of 50–1000 bp (for a 4% gel) but offers improved handling characteristics.

2.3 Procedures

1. Estimate the amount of starting material required to generate
    the desired quantity of the fragment to be isolated. This can be
    calculated using the simple relationship:

    $$N/n \times f = S$$

    where N is the size of the starting material, n is the size of the
    desired fragment, f is the quantity of the desired fragment and S
    is the quantity of the starting material. Thus to obtain 0.5 µg of
    the 23 kb λ HindIII fragment, 1 µg of digested DNA should be
    loaded on the gel. Likewise, 6 µg of digested λ would be
    required to produce 0.5 µg of the 4.4 kb fragment.

2. Where large quantities of DNA are loaded on a gel, several
    tracks must be employed to prevent local overloading. It is
    often convenient to tape together the adjacent teeth of a gel
    comb to produce a long loading well.

3. A thin gel comb should be used to produce tight bands. A 1-mm
    thick comb is ideal for most applications.

4. Choose the correct agarose, at an appropriate concentration, to
    produce maximum resolution of the desired fragment. DNA
    run in LMT gels migrates slightly faster than in standard
    melting temperature agarose.

5. Select an appropriate gel buffer. There is some evidence to
    suggest that TAE is the best buffer to use when DNA is to be
    recovered from a gel. Boron atoms from TBE buffer form
    complexes with agarose [5], making the extraction of DNA
    more difficult. Extraction procedures can be modified when
    TBE is used (see glass powder method). TBE generally shifts
    the separation range of a given agarose ·concentration, so that
    slightly smaller molecules are preferentially separated by its
    improved sieving characteristics. TBE is also desirable when
    gels are to be run at high voltages in order to reduce run times.
    The lower conductivity of this buffer generates less heat than
    TAE.

6. At the end of the run, DNA is detected by staining the gel for
    15 min in ethidium bromide at 1 µg/mL in gel buffer.

7. Detain by soaking the gel in gel buffer for 30 min, with gentle agitation. The inclusion of $MgSO_4$, at 1 mM, will accelerate destaining.
8. DNA destined for critical cloning experiments should not be subjected to ethidium bromide and ultraviolet (u.v.) light. In this situation, cut off the outer tracks of the gel, stain, identify the desired band under u.v. and cut nicks in the gel to mark the position of the band. Re-assemble the gel and cut out the band from the unstained section.
9. Where ethidium bromide and u.v. light are used to detect the DNA band to be excised, use a 300 nm light source and try not to expose the gel to u.v. for more than about 2 min. Exceeding this time leads to progressive damage to the DNA [6].
10. Excess agarose should be trimmed from the excised gel by turning the slice on its side and cutting off non- fluorescent agarose.
11. Use a sterile scalpel or razor blade to carry out excisions.
12. The gel slice is now ready for DNA extraction by any of the methods described in this chapter.

2.4 Technical Notes

(1) The presence of sulphate in agarose is an undesirable characteristic, for its presence in high concentrations is correlated with high EEO, non-specific binding and enzyme inhibition. Sulphate concentrations below 0.15% are associated with high purity agarose.

(2) The electroendosmosis (EEO) of agaroses is a property which affects the electrophoretic mobility of DNA. Agarose contains anionic residues such as ester sulphates and pyruvate. Associated with these anionic residues are hydrated counter cations, which move towards the cathode during electrophoresis. The migration of these cations and associated water of hydration is in the opposite direction to the movement of DNA, which migrates towards the anode. This liquid flow towards the cathode is called electroendosmosis or EEO and tends to slow the migration of DNA moving in the opposite direction. EEO is quoted numerically as relative mobility (–mr). The higher the figure, the higher the EEO and the slower the rate of DNA migration.

(3) Readers dissatisfied with the purity of an agarose batch can attempt to re-purify it using an ion exchange resin [7].

(4) Gel running buffers contain EDTA, which inhibits DNase activity. The concentration of EDTA should be kept to a minimum (1 mM or less), for excessive EDTA inhibits subsequent enzymatic reactions. Alternatively, the sequestering effect of EDTA can be counteracted by adding an equimolar amount of MgCl$_2$.

3.   SPIN-X FILTRATION METHOD

3.1   Principle

Vogelstein [8] described a method for separating DNA from agarose, by centrifugation, using a home-made apparatus. A commercially available filter unit from Costar (Spin-X) (Figure 1) is even more effective than the Vogelstein approach. The attraction of this method is its extreme simplicity. The agarose slice is placed in the Spin-X unit and centrifuged. The cellulose acetate membrane prevents the passage of agarose whilst allowing DNA to pass through. The DNA is

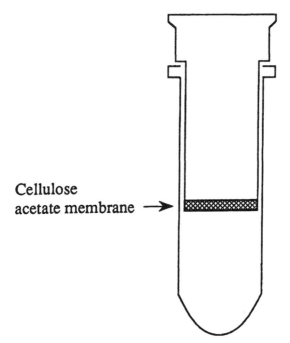

Cellulose
acetate membrane →

Figure 1.   Spin-X centrifuge filter unit.

recovered, sterile, in the gel electrophoresis buffer and can be concentrated by ethanol precipitation. The efficiency of recovery can be increased by a small modification reminiscent of methods published by Tautz and Renz [9] and Koenen [10], in which residual agarose is frozen and then re-centrifuged to extract every last drop of liquid (and DNA). This method provides good recoveries of high quality DNA, capable of undergoing subsequent enzymatic manipulations without further purification.

3.2 Materials

Costar Spin-X centrifuge filter units: 0.22 µm cellulose acetate.

3 M sodium acetate solution pH 5.5.

Ethanol (fresh), stored at –20°C.

70% ethanol, stored at –20°C.

TE (sterile)
   10 mM Tris–HCl
   1 mM EDTA
   pH 7.5.

3.3 Procedure

1. Run the sample in any high quality agarose.
2. Place the excised band (up to 500 mg) in the Spin-X filter cup and macerate, using a sterile implement, taking care not to damage the membrane.
3. Spin for 30 min at room temperature in a microfuge, at full speed.
4. Remove the filter cup and freeze.
5. Replace the frozen filter cup on the collection tube and spin for another 15 min.
6. Measure the volume of the filtrate, with a micropipette, and add 0.1 volumes of 3 M sodium acetate solution pH 5.5, mix, add 2.5 volumes of fresh ethanol, mix and incubate at –20°C for 1 h.
7. Spin for 10 min in a microfuge, at full speed, at 4°C.

8.   Aspirate off the liquid with a fine tip pastette and discard.
9.   Add 100 µL of 70% ethanol and spin for 5 min at 4°C.
10. Aspirate off the liquid with a fine tip pastette and discard.
11. Air dry the tube or evaporate briefly in a freeze drier and take up the DNA pellet in 10–20 µL TE.
12. Run out a small aliquot on a minigel to check for recovery, and store the rest of the sample at –20°C.

3.4   Technical Notes and Troubleshooting

1.   Do not overdry the DNA pellet, for it will be difficult to redissolve.
2.   DNA dissolves readily in slightly alkaline TE.
3.   Ethanol precipitation of small amounts of DNA (<1 µg) can be assisted by the addition of 1 µl of mussel glycogen (10 mg/mL) as carrier [11].

4.   AGARASE METHOD

4.1   Principle

Agarase is an agarose digesting enzyme isolated from *Pseudomonas atlantica*. The enzyme converts molten agarose into neoagaro-oligosaccharides, which on cooling do not solidify and DNA molecules trapped within the agarose are liberated. The carbohydrate reaction products, which do not interfere with most subsequent enzymatic manipulations, are ethanol soluble and so can be separated from the DNA during ethanol precipitation. This method produces a good yield of pure DNA and is particularly suitable for the isolation of very large DNA molecules [12,13].

4.2   Materials

Agarase is available from Boehringer, New England Biolabs and Calbiochem.

Agarase buffer
  10 mM Bis Tris–HCl pH 6.5
  1 mM EDTA

50 mM NaCl

3 M sodium acetate solution pH 5.5.

Fresh ethanol at –20°C.

TE pH 7.5.

10 mM Tris–HCl

1 mM EDTA

4.3 Procedure

1. The DNA should be run in any good quality low melting temperature agarose or FMC NuSieve.
2. Cut the gel slice into small (about 100 µL) blocks and soak in 10 times the volume of agarase buffer, for 1 h, at room temperature, with periodic mixing.
3. Transfer each block to a 1.5 mL. microcentrifuge tube and heat at 68°C for 5 min or until the agarose is molten.
4. Transfer the tube to a water bath maintained at the manufacturer's recommended reaction temperature for agarase (40°C to 45°C) and allow it to equilibrate for 10 min
5. Add 1 unit of agarase per tube, mix gently and incubate for 1 h.
6. Incubate the tube on ice for 10 min to check that the reaction has gone to completion. Check for solidified agarose and if present, re-melt, add additional agarase and re-incubate.
7. Add 0.1 volumes of 3 M sodium acetate solution pH 5.5, mix and maintain on ice for 15 min.
8. Spin in a microcentrifuge for 15 min to pellet any undigested carbohydrates.
9. Transfer the supernatant to another microcentrifuge tube, add 2.5 volumes of fresh ethanol, mix and incubate at –20°C for 1 h.
10. Thereafter, follow the standard ethanol precipitation procedure (see Section 3.3, steps 7–12).

4.4 Technical Notes and Troubleshooting

1. Agarase exhibits maximum activity at pH 6.5. Digestion can take place in electrophoresis buffer by omitting step 4.3.2, but twice as much enzyme is then required.
2. It is important to check that the agarose is completely melted at

step 3 and that no solidification occurs at step 6. This can be achieved by gently drawing up the contents of the tube into a pastette and checking for lumps of agarose.

3. A water bath or hot block can be used for melting the agarose. A particularly versatile and intense source of heat is provided by the 'Intelligent Heating Block' from Cherlyn Electronics Ltd., Kings Hedges Road, Cambridge, CB4 2QH, England. This apparatus takes 1.5 mL microcentrifuge tubes.

4. Incomplete agarose digestion can be caused by a failure to allow the tube to cool to the agarase reaction temperature before adding the enzyme.

5. ELECTROELUTION METHOD

5.1 Principle

Many different electroelution techniques have been devised and most rely on placing a dialysis membrane or a DEAE-filter in an electric field between the gel and the anode and trapping the DNA on the membrane surface (see Chapter 7). This approach can be labour intensive and the recovery can be variable due to the DNA becoming irreversibly bound to the membrane, even following a brief reversal in the polarity of the electric field. A more efficient approach avoids the intervening dialysis membrane and permits migration of DNA directly on to the anode. A brief reversal in polarity then releases the bound DNA from the electrode into the surrounding buffer. The method is similar, in principle, to the simple procedure described by Duro et al. [14] and developed in a commercial apparatus, available from Hoefer (Figure 2). In this device, a cathode is brought in contact with the gel slice, which is surrounded by a small volume of buffer. The anode is immersed in buffer within the collection tube. On application of a d.c. field, as in standard agarose gel electrophoresis, negatively charged DNA molecules, attracted to the anode, migrate from the gel into the surrounding buffer. Tests in my laboratory using this apparatus have shown that DNA, ranging in size from 2–23 kb, is efficiently recovered and in a sufficiently pure form to allow further enzymatic manipulations without extensive additional purification. Any good quality agarose (equivalent to FMC GTG standard) can be used with this method. The cost of purchasing or making this

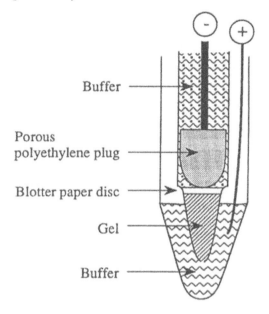

Figure 2.   Diagramatic representation of a Hoefer
electroelution tube.

apparatus can be justified by its use in recovering DNA and proteins
from polyacrylamide gels too.

5.2   Materials

Hoefer GE200 SixPac Gel Eluter. This apparatus allows
simultaneous electroelution of up to six samples and incorporates a
cooling block through which chilled water can be circulated, if
desired. Operating the apparatus in a cold room is easier and prevents
the overheating of thermolabile samples.

TAE buffer × 1 and × 4 (Section 2.2.1)

3 M sodium acetate pH 5.5.

Ethanol

TE pH 7.5.
   10 mM Tris–HCl
   1 mM EDTA

5.3   Procedure

1.   Pipette 80 µL of 1 × TAE into the inner tube.
2.   Cut the gel slice into small pieces and introduce into the tapered end of the inner tube using the packing rod provided. The tube will hold up to 250 µL of agarose.
3.   Eliminate any air bubbles that may have become trapped around the gel slice and add a further 220 µL of 1 × TAE.
4.   Introduce the blotter paper disc into the tube and compress it to pack down the agarose. This is a crucial step, which will prevent buffer leaking from the inner to the outer tube.
5.   Introduce the polyethylene plug into the tube and push down firmly. Eliminate any trapped air bubbles.
6.   Check that the buffer level is about 5 mm below the rim of the tube and adjust if necessary.
7.   Cut the tip off the inner tube to allow electrical continuity between the gel slice and the buffer in the outer tube.
8.   Fit the electrode cap (which carries the cathode and anode) to the inner tube, making sure that the cathode contacts the polyethylene plug.
9.   Pippette 300 µL of 4 × TAE into the outer tube and, maintaining the inner tube in a vertical position, introduce it into the outer tube and connect the electrode cap to the elutor.
10.   Electroelute at 50 V (>0.05 mA) for 2 h for DNA larger than about 3 kb but for only 1 h for molecules smaller than this. Reverse the polarity for 5 s at the end of the run.
11.   Disconnect the electrode cap from the apparatus and withdraw the cap and inner tube, wiping the anode against the inner wall of the outer tube to dislodge any liquid adhering to it.
12.   The eluted DNA can be concentrated by ethanol precipitation (see Section 3.3, steps 6–12).

5.4   Technical Notes and Troubleshooting

1.   It is difficult to see the gel slice through the semi-opaque inner tube wall, so by adding just a portion of the final 300-µL volume of buffer it is possible to see when the tube is full of agarose.
2.   It is essential to eliminate all air bubbles from the apparatus,

for they will result in poor electrical continuity and hence reduce the recovery of DNA. The Hoefer apparatus has indicator lights which can inform the operator of poor continuity.

3. Hoefer do not recommend using FMC Seakem ME or NuSieve GTG agaroses, which give poor recoveries of DNA in their apparatus.

4. The high concentration buffers used in this procedure maintain ionic strength and hence allow rapid elution.

6. GLASS POWDER METHOD

6.1 Principle

Vogelstein and Gillespie [15] first demonstrated the enormous capacity of powdered glass to bind DNA, in the presence of a high concentration of sodium iodide. This chaiotropic salt can also be used to dissolve agarose when applied at final concentrations exceeding 2.5 M. The DNA–glass complex is then treated with more sodium iodide to dissolve any remaining agarose and then washed extensively with Tris buffered ethanol to remove the sodium iodide. The DNA is finally eluted from the glass by treatment with a small volume of low ionic strength buffer. Originally, this method was restricted to agarose gels prepared in TAE, but DNA can now be recovered from gels run in TBE by the addition of mannitol or sodium phosphate to the sodium iodide solution. Commercially available kits are available (Geneclean 11 from Bio101, Vitrogene from Cambio and Qiaex from Diagen), but this protocol describes a method which provides DNA recoveries as good as any kit and at a fraction of the cost.

6.2 Materials

Silica for preparing glass powder: Sigma silica (not fumed) catalogue No. S 5631.

Nitric acid (concentrated).

Sodium iodide solution
    Make up as a 6 M solution in TE pH 7.5 and include 1.5 g sodium

sulphite per 100 mL as an antioxidant. Store in the dark at 4°C.

Mannitol
　　1 M mannitol
　　heat to 50°C to dissolve. Autoclave.

1 M sodium phosphate solution pH 6
　　10 mL of 1M monobasic sodium phosphate (NaH_2PO_4) add
　　sufficient 1 M dibasic sodium phosphate(Na_2HPO_4), usually 4–5
　　mL, until the pH is 6.

Ethanol wash
　　50% ethanol (fresh)
　　0.1 M NaCl
　　10 mM Tris–HCl pH 7.5
　　1 mM EDTA.

TE pH 7.5.
　　10 mM Tris–HCl
　　1 mM EDTA

6.3.1　Preparation of Glass Powder

1. Mix 100 g of silica in 250 mL of distilled water and agitate with a magnetic stirrer for 1 h.
2. Allow the silica to sediment for 90 min at unit gravity.
3. Carefully recover the supernatant, containing the glass powder, taking care not to disturb the sediment.
4. Spin at 13 000 g for 10 min.
5. Resuspend the pellet in 100 mL of distilled water.
6. Add an equal volume of concentrated nitric acid.
7. Heat to boiling point in a fume hood and allow to cool to room temperature.
8. Spin at 13 000 g for 10 min.
9. Resuspend in 200 mL of distilled water.
10. Repeat steps 8 and 9 several times until the pH is 7.
11. Finally store the glass powder as a 50% slurry at −70°C.

6.3.2　Procedure for DNA Recovery

1. Weigh the agarose slice and cut into small pieces.
2. Transfer up to 250 mg to a 1.5 mL microcentrifuge tube.
3. Add 3 volumes of 6 M sodium iodide solution (1 g of agarose

has a volume of 1 mL).

4. If the gel was prepared in TBE buffer, add 0.1 volumes of either 1 M mannitol or 1 M sodium phosphate buffer pH6.
5. Incubate at 55°C for 5–10 min, until the agarose has dissolved. Mix contents periodically to aid dissolution.
6. Add 1 µL of glass powder slurry per 1 µg of DNA and mix thoroughly.
7. Incubate on ice for 10 min, mixing periodically.
8. Spin for 10 s, remove supernatant and save it.
9. Resuspend the pellet in 300 µL of sodium iodide solution.
10. Spin for 10 s.
11. Resuspend the pellet in 300 µL of ethanol wash and spin for 10 s Remove the supernatant.
12. Repeat step 11 twice more and then remove the final wash.
13. Spin the tube containing the pellet for 10 s and remove any residual liquid with a fine tipped pastette.
14. Air dry the pellet and take it up in 20 µL TE.
15. Incubate at 50°C for 10 min.
16. Spin for 1 min and recover the supernatant containing the DNA.
17. Add another 20 µl of TE to the pellet and repeat steps 15 and 16, but keep this batch separate from the earlier elution.
18. Analyse both batches by running small aliquots out on a minigel to determine the efficiency of recovery.

6.4 Technical Notes and Troubleshooting

1. Check very carefully that the agarose has dissolved by slowly drawing the contents of the tube up into a pastette and then ejecting into the tube. Undissolved agarose will be immediately apparent.
2. If the agarose does not dissolve, this is probably caused by too low a final concentration of sodium iodide. Add more and repeat the incubation.
3. Mix the glass powder slurry vigorously immediately before use by vortexing until the suspension is homogeneous.
4. It is wise to save the supernatant from step 8 until the successful recovery of DNA has been demonstrated. If DNA has been lost at this stage it can be recovered by repeating steps 6–8.

5. It is important not to vacuum dry the pellet at step 14. This can lead to difficulties in eluting the DNA.

6. The majority of the DNA should be eluted with the first addition of TE, but a subsequent treatment should elute any remaining.

7. 'IN GEL' LABELLING

7.1   Principle

'In gel' labelling of DNA provides a rapid and economical method for radioactive probe preparation. A low melting temperature agarose gel slice, containing the DNA fragment to be labelled is melted and an aliquot transferred to a fresh tube. Labelling is performed using random priming by adding nucleotides, reaction mixture, 32P label and Klenow enzyme to the denatured DNA aliquot and incubating at 37°C. At this temperature, the agarose remains molten and at the termination of the reaction, unincorporated label can be removed by passing the DNA through a Sephadex G50 column. Whilst it is not so easy to determine the concentration of DNA in a gel slice, the random priming reaction can be performed equally well with DNA amounts ranging from 10 ng to 2 µg.

7.2   Materials

Boehringer random prime labelling kit.

Deoxycytidine 5′–[α^{32}P] triphosphate, 3000 Ci/mmol; Amersham PB10205.

Whatman GF/B 2.4 cm circles.

5% TCA solution

NICK columns - Pharmacia.

TNE
  10 mM Tris pH8
  1 mM EDTA
  200 mM NaCl.

0.2 M EDTA pH8.

7.3 Procedure

1. DNA should be separated in low melting temperature agarose, for example, Seaplaque GTG of Nusieve GTG.
2. The excised gel slice is incubated at 68°C until melted.
3. Remove 11 μL and transfer to a 1.5 mL microcentrifuge tube.
4. Incubate the tube at 95°C for 10 min to denature the DNA.
5. Spin for 2 s and transfer the tube immediately to a 37°C water bath.
6. Immediately add 5 μL of pre-warmed reaction mix/nucleotide cocktail (made up of 1 μL of each of dATP, dGTP, dTTP and 2 μL of reaction mixture).
7. Add 3 μL of α^{32}P dCTP and 1 μL of Klenow enzyme (2 units).
8. Incubate at 37°C for 1 h.
9. Measure the incorporation by removing 1 μL from the tube, mixing with 3 μL of warm (37°C) water and spotting on to a Whatman GF/B filter.
10. Measure the total radioactivity by Cerenkov counting in a scintillation counter.
11. Pass about 20 mL of 5% TCA through the filter and count again.
12. Express the counts obtained in step 11 as a percentage of the figure obtained in step 10; this is the percentage incorporation, which should be at least 50%.
13. Stop the reaction by adding 2 μL of 0.2 M EDTA pH8 and heating to 65°C for 10 min.
14. Remove the unincorporated label by adding 80 μL of pre-warmed (37°C) TNE buffer to the reaction and loading on to a NICK column, which has been prepared by decanting off the storage liquid and rinsing the column with 10 mL of TNE.
15. Allow the liquid to run through the column and then wash with 300 μL of TNE. Discard the run-through and wash.
16. Elute the labelled DNA with 400 μL of TNE, which is collected in a microcentrifuge tube.
17. Store the eluted DNA at −20°C or use it immediately for hybridisation.

7.4   Technical Notes and Troubleshooting

1. From steps 4 to 14 it is essential that the agarose is kept molten by not allowing the tube to cool below 37°C and adding only pre-warmed reagents.
2. If the original gel slice contained 0.8% agarose, then this concentration would be reduced to 0.4% in the random prime reaction. It is known that many enzymes work best at agarose concentrations of 0.5% or less.
3. By the time the reaction mixture is loaded on to the NICK column, the agarose will be reduced to <0.1%, so that no solidification occurs in the column.
4. If the incorporation of label is less than 50%, add another 0.5 µL of Klenow and continue incubation for 30–60 min.
5. Following denaturation of the DNA, the addition of the other reagents, required for the labelling reaction, should be completed as quickly as possible, for during this time the DNA will be re-annealing.
6. The DNA can be labelled non-radioactively by following this protocol but using digoxigenin-11-dUTP (Boehringer) instead of the radioactive isotope, incorporating cold dATP, dGTP and dCTP and, of course, omitting steps 9–12.
7. The gel should be prepared in a buffer containing a low EDTA concentration, preferably TAE. High EDTA concentrations will inhibit the labelling reaction.

ACKNOWLEDGMENT

I would like to thank Douglas Stuart for preparing the figures used in this chapter.

REFERENCES

[1]   Struhl, K. A rapid method for creating recombinant DNA molecules. Biotechniques, 3 (1985) 452–453.
[2]   Parker, R.C. and Seed, B. Two-dimensional agarose gel electrophoresis 'SeaPlaque' agarose dimension. Methods Enzymol., 65 (1980) 358–363.
[3]   Maniatis, T., Fritsch, E.F. and Sambrook, J. *Molecular Cloning. A Laboratory*

Manual. 2nd. edn., Cold Spring Harbor Laboratory, Cold Spring Harbor, N. Y., 1989.

[4] Smith, H.O. Recovery of DNA from gels. Methods Enzymol., 65 (1980) 371–380.

[5] Peats, S., Nochumson, S. and Kirkpatrick, F.H. Effects of borate on agarose gel structure. Biophys.J., 49 (1986) 91a.

[6] Dumais M.M. and Nochumson, S. Small DNA fragment separation and M13 cloning directly in melted NuSieve GTG agarose gels. Biotechniques 5 (1987) 62–67.

[7] Arveiler, B., Vincent, A. and Mandel, J.-L. Toward a physical map of the Xq28 region in man: linking color vision, G6PD and coagulation factor V111 genes to an X-Y homology region. Genomics 4 (1989) 460–471.

[8] Vogelstein, B. Rapid purification of DNA from agarose gels by centrifugation through a disposable column. Anal. Biochem. 160 (1987) 115–118.

[9] Tautz, D. and Renz, M. An optimized freeze-squeeze method for the recovery of DNA fragments from agarose gel. Anal. Biochem., 132 (1983) 14–19.

[10] Koenen, M. Recovery of DNA from agarose gels using liquid nitrogen. Trends Genet., 5 (1989) 137.

[11] Helms, C., Graham, M.Y., Dutckik, J.E. and Olson, M.V. A new method for purifying lambda DNA from phage lysates. DNA 4 (1985) 39–49.

[12] Anand, R., Villasante, A. and Tyler-Smith, C. Construction of yeast artificial chromosome libraries with large inserts using fractionation by pulsed field gel electrophoresis. Nucl. Acids Res., 9 (1989) 3425–3433.

[13] Albertsen, H.M., Abderrahim, H., Cann, H.M., Dausset, J., Le Paslier, D. and Cohen, D. Construction and characterization of a yeast artificial chromosome library containing seven haploid human genome equivalents. Proc. Natl. Acad. Sci. USA, 87 (1990) 4256–4260.

[14] Duro, G., Izzo, V., Barbieri, R., Cantone, M., Costa, M.A. and Giudice, G. A method for eluting DNA in a wide range of molecular weights from agarose gels. Anal. Biochem., 195 (1991) 111–115.

[15] Vogelstein, B. and Gillespie, D. Preparative and analytical purification of DNA from agarose. Proc. Natl. Acad. Sci. USA, 76 (1979) 615–619.

Chapter 4

PLASMID ISOLATION PROCEDURES

A. I. Knight

OUTLINE

Methods in Gene Technology, Volume 2, pages 65–73
Copyright © 1994 JAI Press Ltd
All rights of reproduction in any form reserved.
ISBN: 1-55938-264-3

1.   CAESIUM CHLORIDE GRADIENTS

Caesium chloride gradient centrifugation separates DNA molecules on the basis of density. Displacement of DNA strands by intercalating agents such as ethidium bromide reduces the density of double stranded DNA (see Chapter 1). Covalently closed circular plasmid molecules bind less dye due to their supercoiled structure and therefore band at a higher density in Caesium chloride gradients [1]. Although many techniques are available for the rapid and simple purification of plasmid DNA, caesium chloride gradients remain useful for a number of reasons. Firstly, although alternative plasmid isolation techniques for *Escherichia coli* are well developed, they are often inefficient for other organisms. Secondly although these alternative procedures are effective for high copy number cloning vectors they are not as useful for low copy number and high molecular weight plasmids where the yields of plasmid DNA are low. Thirdly large scale preparations using this technique are a useful means of providing stocks of highly purified DNA of commonly used laboratory cloning vectors which can last many years.

The isolation of plasmids using this technique requires a crude plasmid isolation procedure such as a simple cleared lysate method in which the cells are gently lysed by a combination of lysozyme and detergent treatments and the bulk of the chromosomal DNA is pelleted by centrifugation. Although the chromosomal DNA will be fragmented to some extent during lysis, the success of the clearing spin relies on the absence of gross shearing, as high molecular weight DNA will co-precipitate with the cell debris. It is preferable *not* to RNase treat the extract at this stage, as the RNA fragments generated will contaminate the plasmid DNA.

Caesium chloride and ethidium bromide are then added to the supernatant (cleared lysate) and the resulting solution subjected to ultracentrifugation. The resulting plasmid band can be visualised under ultraviolet light and removed by use of a needle and syringe. Plasmid DNA is finally recovered by extracting the ethidium bromide with isopropanol and precipitating the plasmid DNA from the caesium chloride solution.

1.1 Basic Cleared Lysate Procedure

1. Grow 150 mL of culture overnight in the presence of appropriate antibiotics for plasmid selection.
2. Harvest the cells by centrifugation at 3000 g for 10 min at 4°C.
3. Resuspend the cell pellet in 3.5 mL of 25% sucrose in 50 mM Tris-Cl, pH 8.0.
4. Add 0.5 mL lysozyme (20 mg/mL in 0.25 M EDTA pH 8.0, prepared immediately before use). Leave on ice for 10 min.
5. Add 3.5 mL of 0.25 M EDTA pH 8.0. Leave on ice for 5 min.
6. Add 5 mL Brij–Doc (1% Brij 58, 0.4% deoxycholate in 10 mM Tris-Cl, 1 mM EDTA pH8.0). Lyse cells by gentle pipetting. Leave on ice for 30 min.
7. Centrifuge at 15 000 r.p.m. for 45 min at 4°C.
8. Carefully remove 10 mL of the resulting supernatant (cleared lysate) for caesium chloride density gradient centrifugation. It is important to ensure that the pellet is not disturbed at this stage.

1.2 Gradient Preparation

1. Add 4.75 g of caesium chloride and 150 µL of ethidium bromide solution (1.5% in H_2O) to 5 mL aliquots of cleared lysate. This gives a final density of 1.55 g/mL.
   CAUTION: ethidium bromide is mutagenic and carcinogenic; wear gloves when using it. Pellets of ethidium bromide are commercially available for preparation of solutions.
2. Load the gradient solution into Beckman Quickseal tubes and top up the tubes with liquid paraffin.
3. Ensure that the tubes are balanced, heat seal, and centrifuge for 15 h at 150 000 g at 15°C (Beckman 70.1 rotor). Centrifugation time may be reduced to 6 h by centrifugation at 220 000 g using a vertical rotor (Beckman Vti.65).

1.3 Recovery of Plasmid DNA

After centrifugation, carefully remove the tubes from the centrifuge, taking care not to disturb the plasmid band. Firmly clamp a centrifuge tube in a retort stand and illuminate with u.v. light (wear a u.v. protective mask). Two fluorescent DNA bands are frequently seen: a faint upper band (which may even be absent if the clearing spin is very effective), corresponding to chromosomal DNA, and a lower (usually stronger) band corresponding to supercoiled plasmid DNA. The relative amounts of the two bands will depend on the copy number of the plasmid and the efficiency with which chromosomal DNA is removed in the clearing spin. RNA will form a diffuse band at the bottom of the tube, and denatured protein may be present as a bright red layer at the top.

Wearing gloves and taking care to avoid needlestick injuries, use a 21 g needle to puncture a hole in the top of the tube. Connect a fresh 21 g needle to a 1 mL syringe and ensure the syringe plunger moves freely. Insert the syringe needle 1 mm below the plasmid band and approximately half way into the tube. Slowly withdraw the plasmid band with the syringe and at the same time gently move the syringe needle from side to side. With experience the plasmid DNA is usually recovered in approximately 0.6 mL.

1.4   Extraction of Ethidium Bromide

Before precipitating the DNA from the material recovered from the gradient, it is necessary to remove the ethidium bromide by extracting with isopropanol saturated with caesium chloride solution (ICC).

ICC is prepared by adding 10 g of caesium chloride to 10 mL of TE buffer (10 mM Tris-Cl, 1 mM EDTA, pH 8). 80 mL of isopropanol is then added slowly and the solution shaken to dissolve any precipitating caesium chloride. The upper isopropanol phase (ICC) is then used for removal of ethidium bromide.

The gradient extract is mixed with a half volume of ICC and allowed to settle. The upper phase containing the red ethidium bromide is removed and the extraction repeated until all visible traces of ethidium bromide have been removed (usually three times).

1.5   Precipitation of Plasmid DNA

1. Add 75 µL of 3 M sodium acetate pH 5.2 and 425 µL of H_2O to 300 µl of purified gradient extract.
2. Add 750 µL of isopropanol (*not* ICC) and mix well. Allow the precipitate to form at room temperature for 10 min. Centrifuge

(using a microcentrifuge at 10 000 *g*) for 10 min.

3. Carefully remove all the supernatant and resuspend the plasmid DNA pellet in 200 µL of TE buffer. Re-precipitate the DNA by adding 20 µL of 3 M sodium acetate pH 5.2 and 440 µL of cold (−20°C) ethanol; leave the tube at −70°C for 30 min. Centrifuge as above for 10 min.

4. Wash the pellet by the addition of 500 µL of 70% ethanol (*not* absolute ethanol), centrifuge for 10 min, dry the pellet briefly under vacuum and resuspend the plasmid DNA in TE buffer.

5. Finally determine the DNA concentration and store at −20°C.

1.6 Comments

Care must be taken during this procedure to avoid exposure to u.v. light and needlestick injuries. Gloves should be worn at all times when handling ethidium bromide which is a powerful mutagen.

Although the method described here is relatively slow compared with other methods (see Section 2, and Chapter 5) it provides an efficient method for the recovery of good yields of highly purified plasmid DNA suitable for all genetic manipulations.

Perhaps the most frequently encountered problem with this method lies in the removal of the plasmid band from the gradient by syringe. This can be technically demanding especially when dealing with low copy number plasmids and usually requires some practice to obtain efficient recovery from the gradients.

The cleared lysate procedure may not give good yields with large plasmids, which can be precipitated with the chromosomal DNA. If the clearing spin is omitted, the resulting material is too viscous to be separated effectively by gradient centrifugation. However, the viscosity can be reduced by controlled shearing of the DNA, either by vortexing (see Chapter 6) or by repeated passage through a syringe needle. The supercoiled plasmid DNA is more resistant to shearing than is the chromosomal DNA. If total DNA is subjected to caesium chloride gradient centrifugation, the (upper) chromosomal band will be very dense, and it may be advantageous to remove this before attempting to recover the plasmid DNA. Alternative methods for the purification of large plasmids are outlined in Chapter 6.

2. RAPID SMALL-SCALE METHODS

Minipreparations of plasmid DNA represent an invaluable technique
for analysis of recombinant molecules. Two principal methods are
widely used for the analysis of plasmid DNA from *E. coli*. The
boiling method of Holmes and Quigley [2] and alkaline lysis
procedures [3,4] both provide rapid and simple methods of extracting
plasmid DNA for analysis and genetic manipulations. In our hands
the alkaline lysis method although slower than the boiling method
provides better quality DNA preparations. The principle of these
techniques relies on the greater resistance to alkali or heat
denaturation of covalently closed circular plasmid molecules
compared with chromosomal DNA (see also Chapter 5).

2.1   Alkaline Lysis Method

2.1.1   Reagents

Solution I
    50 mM glucose
    25 mM Tris-Cl (pH 8.0)
    10 mM EDTA (pH8.0)

Solution II (Freshly prepared solution)
    0.2 M NaOH
    1% SDS

Solution III
    5 M potassium acetate   60.0 mL
    Glacial acetic acid   11.5 mL
    H_2O 28.5 mL

2.1.2   Procedure

1. Inoculate 2 mL of broth containing appropriate antibiotics to
    select for plasmid markers and incubate overnight. Harvest the
    cells from 1.5 mL of this culture by centrifugation at 12 000 g
    for 30 s in a microcentrifuge.
2. Remove the supernatant by aspiration and resuspend the pellet
    in 150 µL of ice cold solution I by vortexing.

3. Add 300 μL of freshly prepared solution II and mix by gentle inversion. (The solution should become clear as the cells lyse.) Leave the tubes on ice for 5 min.
4. Add 225 μL ice cold solution III. Vortex the tubes gently for 10 s and then leave on ice for 5 min.
5. Centrifuge at 12 000 g for 10 min at 4°C.
6. Add an equal volume of phenol–chloroform, equilibrated with buffer at pH 8.0 (see Chapter 1) to 550 μL of the supernatant from step 5. Vortex briefly and centrifuge at 12 000 g for 5 min in a microfuge.
7. Carefully remove 450 μL of the upper aqueous phase and add 900 μl of ethanol, at room temperature. Mix well and allow the DNA to precipitate at room temperature for 2 min. Centrifuge at 12 000 g for 5 min to pellet the DNA.
8. Wash the plasmid DNA pellet with 1 mL of 70% ethanol and centrifuge at 12 000 g for 5 min at 4°C. Very carefully remove the supernatant and dry the pellet under vacuum for 2 min.
9. Resuspend the pellet in 30 μl of TE buffer containing 20 μg/mL DNAase free RNAase.
10. Use 1–2 μL for agarose gel electrophoresis and restriction digests. Typical yields for high copy number plasmids such as pUC plasmids are 3 μg/mL of original culture.

2.2   Boiling Method

2.2.1   Reagents

STET
   8% sucrose
   0.5% Triton X-100
   50 mM EDTA (pH 8.0)
   10 mM Tris HCl (pH 8.0)

TE buffer
   10 mM Tris-Cl pH 8.0
   1 mM EDTA

Lysozyme
   10 mg/mL in TE buffer
   Prepare immediately before use.

sodium acetate
  2.5 M (pH 5.2)

isopropanol

RNase
  20 µg/mL of DNase-free RNase in TE buffer

2.2.2   Procedure

1. Inoculate 2 mL of broth containing appropriate antibiotics to select for plasmid markers and incubate overnight. Harvest the cells from 1.5 mL of this culture by centrifugation at 12 000 g for 30 s in a microcentrifuge.

2. Resuspend the cell pellet in 350 µL STET and add 25 µL of lysozyme solution (10 mg/mL in TE). Vortex briefly.

3. Place the tube in a boiling water bath for 40 s.

4. Centrifuge at 12 000 g for 5 min at 4°C and remove the cell pellet with a toothpick.

5. To the supernatant from step 4, add 40 µL of 2.5 M sodium acetate (pH 5.2) and 420 µL of isopropanol. Vortex briefly and leave at room temperature for 5 min to precipitate the plasmid DNA.

6. Centrifuge at 12 000 g for 5 min at 4°C.

7. Wash the plasmid DNA pellet with 1 mL of 70% ethanol and centrifuge at 12 000 g for 5 min at 4°C. Very carefully remove the supernatant and dry the pellet under vacuum for 2 min.

8. Resuspend the pellet in 30 µL of TE containing 20 µg/mL of DNAase-free RNAase.

9. Use 1–2 µL for agarose gel electrophoresis and restriction digests. Typical yields for high copy number plasmids such as pUC plasmids are 3 µg per mL of original culture.

2.3   Comments

The most commonly encountered problem with 'miniprep' DNA is that although the DNA may cleave extremely well with some restriction enzymes, others may only yield partial digestion. This is usually due to contaminants in the miniprep DNA and is frequently associated with inefficient removal of supernatants from centrifugation steps or with prolonged ethanol precipitations. Where such problems exist they may be overcome by:

(a) performing restriction digests in a large volume (100–200 μl) with a 10 fold excess of enzyme;

(b) re-precipitating the DNA, ensuring that all the supernatant is removed, and washing carefully with 70% ethanol; and

(c) adding (or repeating) an extraction step with phenol– chloroform. Note that small-scale rapid plasmid preparations are usually not stable unless phenol or phenol–chloroform extracted, presumably due to nuclease contamination.

REFERENCES

[1]   Radloff, R., Bauer, W. and Vinograd, J. A dye-buoyant density method for the detection and isolation of closed circular duplex DNA: The closed circular DNA in HeLa cells. Proc. Natl. Acad Sci., 57 (1967) 1514–1521.

[2]   Holmes, D.S. and Quigley, M. A rapid boiling method for the preparation of bacterial plasmids. Anal Biochem., 114 (1981) 193–197.

[3]   Birnboim, H.C. and Doly, J. A rapid alkaline extraction procedure for screening recombinant plasmid DNA. Nucleic Acids Res., 7 (1979) 1513–1523.

[4]   Ish-Horowicz, D. and Burke, J. Rapid and efficient cosmid cloning. Nucleic Acids Res., 9 (1981) 2989–2998.

Chapter 5

PLASMID ISOLATION
COLUMN METHODS

T.J. Hellyer

OUTLINE

Methods in Gene Technology, Volume 2, pages 75–86
Copyright © 1994 JAI Press Ltd
All rights of reproduction in any form reserved.
ISBN:1-55938-264-3

1. INTRODUCTION

Many procedures have been described for the isolation of plasmids which exploit the stability of their supercoiled structure and their small size relative to the bacterial chromosome [1–3] (see also Chapter 4). Highly purified plasmid DNA free of contaminating protein, RNA and chromosomal fragments is frequently required for restriction endonuclease analysis, sequencing reactions and other common manipulations. Caesium chloride-ethidium bromide dye-buoyant density centrifugation (see Chapter 4) has long been the method of choice for the preparative isolation of purified plasmid DNA. However, this technique necessitates ultracentrifugation for an extended period and the use of large quantities of expensive carcinogenic reagents. Column chromatography using a variety of commercially available resins is now a popular alternative for plasmid purification. Chromatographic procedures can be performed rapidly, require no specialized equipment and eliminate the use of toxic substances such as phenol, chloroform and ethidium bromide. Chromatography is particularly useful for the purification of plasmid DNA on a small scale starting with as little as 1 mL bacterial culture.

An exhaustive account providing details of all the available chromatography resins and the manufacturers' recommendations for their use is beyond the scope of this text. However, the principles underlying the most widely applied techniques of anion exchange and gel filtration chromatography are outlined together with a general procedure for the extraction of plasmid DNA suitable for further purification by column methods.

2. PLASMID EXTRACTION

2.1 General Points

Prior to chromatographic fractionation, plasmids must be released in an intact form from the bacterial cell. This is usually accomplished by an alkaline lysis procedure during which the majority of contaminating cellular components are also removed [1,3] (see Chapter 4). The procedure outlined below avoids the use of organic solvents which may have a deleterious effect on the chromatography matrices used in subsequent purification.

The technique entails gentle lysis of the bacteria with detergent followed by selective denaturation of linearized chromosomal DNA at high pH (12–12.5) [1]. Neutralization of the lysate by addition of acidified potassium acetate precipitates chromosomal DNA, protein-detergent complexes and high molecular weight RNA in an insoluble clot which is removed by centrifugation. The resulting lysate is suitable for further purification by either anion exchange or gel filtration chromatography. The finite capacity of both types of purification column requires that considerable care is taken during plasmid extraction in order to minimize contamination of the lysate with cellular debris.

2.2   Reagents

Cell Buffer (**A**):
   50 mM Tris-HCl
   10 mM EDTA; pH 8.0
   RNAase A 100 μg/mL (DNase-free)

Lysis Buffer (**B**):
   Sodium dodecyl sulphate (SDS) 1% (w/v)
   200 mM NaOH

Neutralizing Buffer (**C**):
   2.55 M potassium acetate; pH 4.8

Notes: DNAase-free RNAase is prepared by heating a solution of RNAase (10 mg/mL in 10 mM Tris-HCl; 15 mM NaCl; pH 7.5) to 100°C for 15 min [4]. The enzyme should then be stored in aliquots at –20°C.

Storage Conditions: Cell Buffer **A** is stored at 4°C, buffers **B** and **C** should be kept at room temperature. SDS in Lysis Buffer **B** may precipitate during storage and should be redissolved by warming to 37°C before use.

2.3   Methodology

1.   Harvest the cells from 1–3 mL of an overnight culture of *Escherichia coli* containing not more than 10^9 cells/mL.
2.   Resuspend the bacterial pellet in 300 μL Cell Buffer **A**.
3.   Add 300 μL Lysis Buffer **B** and mix gently by inverting the tube 5-6 times. Incubate at room temperature for 5 min.

4. Add 300 µL neutralizing Buffer **C** and mix thoroughly as in (3). Centrifuge at maximum speed in a microfuge for 15 min.

5. Carefully remove the supernatant avoiding the pellet of cell debris.

2.4 Considerations

1. This procedure is optimized for the recovery of up to about 10 µg of high copy number plasmids such as pUC18/19 or pBluescript from *E. coli*. For plasmids of lower copy number the starting volume of culture should be adjusted appropriately in order to maximize yield. However, both plasmid extraction and column purification are more efficient at low cell/DNA concentration and it is important to avoid an excess of cells. High cell density in the initial suspension produces a viscous lysate which is very difficult to mix. In turn this leads to failure to precipitate all the chromosomal DNA in step (4) and contamination of the sample with cellular debris.

2. Gram-positive organisms are frequently resistant to lysis. In some cases this may be overcome by incorporation in the growth medium of antibiotics with specific activity against the cell wall [5]. Incubation of cells in lysozyme (lysozyme 2 mg/mL in sucrose 0.3 M; 25 mM EDTA; 25 mM Tris-HCl; pH 8.0) before resuspending in Cell Buffer **A** may also enhance lysis.

3. On addition of buffer **B** the cells are rapidly lysed and the suspension becomes clear. Thorough mixing is essential for efficient lysis and denaturation, although excessive vigour should be avoided to prevent fragmentation of chromosomal DNA and damage to the plasmids.

4. Addition of Neutralizing Buffer **C** precipitates the cellular debris as a viscous clot. Again avoid excessive mixing and release of debris into the solution.

5. Centrifugation should leave a clear supernatant although this is frequently contaminated with flakes of potassium dodecyl sulphate. If present in large amounts these may block the chromatography column and should be removed by filtration through a clean nylon mesh.

6. Ethanol or isopropanol precipitation of the lysate at this stage

yields plasmid DNA suitable for many applications. To avoid unnecessary expense it is worthwhile checking whether the desired experiment can be performed with such a crude preparation before undertaking further purification.

3. PLASMID PURIFICATION

3.1 Anion Exchange Chromatography

3.1.1 General Points

Purification of nucleic acids by ion-exchange chromatography is based upon the electrostatic interaction between negatively charged phosphate groups on the polynucleotides and anion exchangers which are immobilized on the column matrix. Once bound to the resin, nucleic acid can be eluted by increasing the ionic strength of the chromatography resin above that required for binding. Proteins, oligomers, RNA and single and double stranded DNA show varying affinities allowing fractionation of the sample under carefully controlled conditions of salt and pH (Table 1).

Table 1. Concentration of NaCl required for elution of nucleic acids and protein from Qiagen anion exchange columns at pH 7.0 in buffer containing 50 mM MOPS; 15% (v/v) ethanol

	NaCl (mM)
Nucleotides	200
Oligonucleotides (10–18 bases)	250–600
Proteins	350
Double-stranded DNA (>10 base pairs)	450–550
Transfer RNA	550–650
Ribosomal RNA	900–1000
Single-stranded DNA	1200–1300
Double-stranded DNA (>150 base pairs)	1350–1450

The following procedure refers specifically to Qiagen Tip-20 anion exchange columns (Diagen GmbH) which are designed for the recovery of up to 10 μg of purified plasmid DNA in a single procedure. The principles involved in the use of larger Qiagen columns for the isolation of up to 500 μg DNA and those of other manufacturers (e.g. Promega Magic Columns) are similar but for

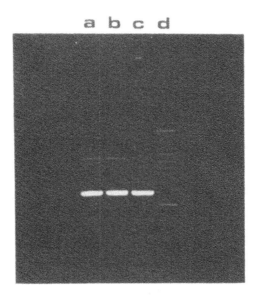

Figure 1.   Ethidium bromide-stained agarose gel showing pBluescript plasmid DNA purified by anion exchange and gel filtration chromatography. Migration was from top to bottom. Loading in lanes **a** and **b** represents 5% total plasmid DNA recovered from 1-mL *E. coli* culture.

Sources of DNA: **a** Qiagen-Tip 20 anion exchange column; **b** Pharmacia Sephacryl S-400 Miniprep Spun-Column; **c** 0.5 µg of commercially prepared pBluescript DNA (Stratagene); **d** 1 kb linear DNA ladder.

detailed information the reader should refer to the companies' own literature.

Figure 1 illustrates the purity of a sample of pBluescript plasmid DNA isolated from *E. coli* according to the above alkaline lysis protocol and fractionated using a Qiagen-Tip 20 anion exchange column.

3.1.2   Reagents

Equilibration Buffer (**I**):
  750 mM NaCl
  50 mM 3-*N*-morpholino-propanesulfonic acid (MOPS)
  15% (v/v) ethanol
  0.15% (v/v) Triton X-100 (optional)
  adjust pH to 7.0

Washing Buffer (**II**):
  1.0 M NaCl
  50 mM MOPS
  15% (v/v) ethanol
  adjust pH to 7.0

Elution Buffer (**III**):
  1.2 M NaCl
  50 mM MOPS
  15% (v/v) ethanol
  adjust pH to 8.0

Isopropanol

70% (v/v) ethanol

Tris-EDTA (TE) Buffer
  10 mM Tris-HCl
  1 mM EDTA
  adjust pH to 8.0

Notes: The pH values of the buffers should be checked regularly since the interaction between nucleic acids and the anion exchange resin is strongly pH dependent. Incorrect buffer formulation is a frequent cause of low yield and/or contamination of the final sample.

Buffers may be prepared from stock solutions of 5 M NaCl and 1 M MOPS in sterile distilled water. Concentrated solutions of MOPS may show a yellow discolouration after prolonged storage but this does not affect buffering capacity.

Storage Conditions: Store all buffers at room temperature.

3.1.3 Methodology

1. Equilibrate a Qiagen-Tip 20 with 1–2 mL buffer **I** either by gravity flow or using a 2 mL syringe.
2. Load the cleared lysate from the above plasmid extraction protocol onto the column.
3. Wash the column with 2 mL buffer **II** and elute the purified DNA with 800 µL buffer **III**.
4. Precipitate the DNA with 0.5–1.0 volumes of isopropanol at room temperature and centrifuge at maximum speed in a microfuge for 30 min either at room temperature or at 4°C.
5. Decant the supernatant and wash the DNA pellet with ethanol

70% (v/v) and re-centrifuge. Dry the pellet briefly in air and resuspend in an appropriate volume of TE buffer (usually 20–30 µL).

3.1.4   Considerations

1. Slow passage of the sample through the column is essential to allow adequate time for binding of the DNA. A flow rate of 1–3 drops/s (maximum of 0.5 mL/min) is optimal throughout the procedure. Gravity flow may used but does not always provide an adequate flow rate unless detergent (Triton X-100) is included in the Equilibration Buffer (I). Alternatively the sample may be forced through using a pipette or a 2 mL syringe fitted with an appropriate adaptor.

2. If the final DNA preparation is found to be contaminated with RNA check the pH of the Washing Buffer (II). If contamination persists, the ionic strength of this buffer may be increased (1.2 M NaCl; 50 mM MOPS; 15% (v/v) ethanol; pH 7.0).

3. Co-precipitation of salt is avoided by carrying out the iso-propanol precipitation at room temperature. Plasmid DNA may be difficult to redissolve if it is allowed to dry completely after washing in ethanol 70% (v/v). If possible leave the final pellet in TE buffer on ice or at 4°C for several hours to dissolve.

4. Confirm the presence of plasmid DNA by running a small sample (1–5 µL) of the final solution on an agarose gel and staining with ethidium bromide. It is advisable to retain all the fractions from the above procedures as well as the column itself until the presence of plasmid DNA in the eluate is confirmed.

5. Anion exchange columns should not be re-used for preparation of different plasmids but may be re-used for the same sample after re-equilibration with buffer I.

3.2   Gel Filtration Chromatography

3.2.1   General Points

Gel filtration or size exclusion chromatography is a method of separating molecules on the basis of size [6]. Columns are prepared from porous spherical beads which are saturated with solvent. The gel particles absorb liquid according to the degree of internal cross-linking to form pores of uniform dimensions. When a mixture of molecules is applied to a column packed with solvent-saturated beads, small molecules are able to penetrate the pores within the gel to differing degrees depending on their size whereas larger molecules are excluded and pass through the column in the liquid phase. Molecules are thus delayed in their passage through the gel in inverse relation to their size. In contrast to ion exchange chromatography, gel filtration requires just a single buffer since the elutant plays no active role in separation, merely serving to carry the molecules through the column.

There is a vast array of gel filtration media which differ in pore size and physical properties. For separation of plasmid DNA from contaminating RNA and proteins, Sephacryl S-400 or S-1000 (Pharmacia) and Bio-Gel A-50m or A-150m (Bio-Rad) are commonly used. Although it is possible to prepare gel filtration columns from these media in the laboratory [7] several pre-packed matrices are available which obviate this time consuming process. Both Pharmacia and Clontech Laboratories produce gel filtration matrices specifically for the rapid purification of mini-preparations of plasmid DNA using spun-column chromatography. This technique involves centrifugation of the sample through a packed column of gel whereby small contaminating molecules are retained within the matrix and purified DNA is recovered in the eluate. Spun-column methods are extremely rapid and have the advantage that several samples may be handled simultaneously.

The following protocol refers to Pharmacia Miniprep Spun Columns although the manipulations involved are similar to those for Clontech Chroma-Spin Columns. The Miniprep gel filtration system comprises pre-packed columns of Sephacryl S-400 which is a gel of cross-linked allyl dextran and N, N'-methylenebisacrylamide with an exclusion limit in the order of 1000 base pairs of DNA [6]. Figure 1 includes an illustration of pBluescript plasmid DNA purified by gel filtration through a Sephacryl S-400 spun-column using the procedure outlined below.

3.2.2   Reagents

Tris-EDTA (TE) Buffer:
  10 mM Tris-HCl
  1 mM EDTA
  adjust pH to 8.0

Isopropanol

3.2.3   Methodology

1. Precipitate the DNA from the cleared lysate obtained above by adding an equal volume of isopropanol and incubating at room temperature for 5 min.
2. Centrifuge at maximum speed in a microfuge for 15–30 min at room temperature or 4°C.
3. Decant the supernatant and invert the tube to drain completely. Dissolve the DNA pellet in a maximum of 50 μL TE buffer.
4. Invert the Miniprep Spun Column several times to resuspend the gel then allow to settle by standing vertically in a rack or clamping in a retort stand.
5. Remove the **top** and **bottom** caps in that order and allow the column to drain by gravity flow.
6. Layer 2 mL TE buffer on top of the gel and allow to drain. Repeat with another 2 mL TE buffer.
7. Place the column in a 15 mL centrifuge tube and centrifuge at 400 g for 2 min in a swing-out rotor.
8. Place a volume of TE buffer equivalent to the volume of DNA sample to be purified on top of the compacted gel and re-centrifuge as in (7).
9. Transfer the column to a clean 15 mL centrifuge tube and apply the DNA sample to the top of the gel.
10. Centrifuge at 400g for 2 min.
11. Recover the sample from the bottom of the centrifuge tube and confirm the presence of plasmid DNA by agarose gel electrophoresis.

3.2.4   Considerations

1. The final sample volume must not exceed 50 μL. Low molecular weight contaminants may elute with the DNA if the sample is applied in a larger volume.

2. A fixed angle rotor should not be used since this enables the sample to run down the side of the column without penetrating the gel matrix and reduces the efficiency of fractionation.

3. The speed of centrifugation is critical. Centrifugation in excess of 400 g or for longer than 2 min may dehydrate the column and cause cracking of the gel bed. This will enable the sample to leak through the column and result in incomplete removal of contaminating RNA and protein. If the gel is too dry DNA may also adhere to the matrix, resulting in low yield of purified plasmid.

   The desired centrifugation speed should be calculated according to the following formula:

$$\text{r.p.m.} = 1000 \times (\text{r.c.f.}/1.12\,R)^{1/2}$$

Where:

   r.p.m. = number of revolutions per minute
   r.c.f. = relative centrifugal force (g), and
   R = distance (mm) from spindle to bottom of rotor bucket

   Strict attention should be paid to centrifugation specifications since these may vary between manufacturers (e.g. Clontech Chroma-Spin Columns require 700 g).

4. The gel appears compacted after centrifugation and will have retracted slightly from the wall of the column. Ensure that the DNA sample is loaded on to the top of the gel and does not run down the sides as this will impair separation.

5. To assist with gel loading, bromophenol blue 0.01% (w/v) may be included in the TE buffer. The dye molecules will be removed during the purification process.

6. Do not overload the column with more DNA than can be obtained from 3 mL of culture as this will lead to poor resolution. If necessary estimate the concentration of DNA in the unpurified sample by comparison with standards of known concentration in an ethidium bromide-stained agarose gel.

7. Gel filtration columns should not be re-used since contaminants retained within the gel are not physically bound to the matrix and may elute with the DNA if another sample is applied.

ACKNOWLEDGMENT

I would like to thank Dr Ian S. Bevan and Dr Charles W. Penn for their invaluable help in preparing this manuscript.

REFERENCES

[1] Birnboim H.C., Doly J. A rapid alkaline extraction procedure for screening recombinant plasmid DNA. Nucl. Acids Res., 7 (1979) 1513–1523.

[2] Kado C.I., Liu S.-T. Rapid procedure for detection and isolation of large and small plasmids. J. Bacteriol., 145 (1981) 1365–1373.

[3] Sambrook J., Fritsch E.F., Maniatis T. *Molecular Cloning: A Laboratory Manual,* 2nd edn., Cold Spring Harbor Laboratory, Cold Springs Harbour, N.Y., 1989, pp. 1.21–1.52.

[4] Sambrook J., Fritsch E.F., Maniatis T. *Molecular Cloning: A Laboratory Manual* 2nd edn., Cold Spring Harbor Laboratory, Cold Spring Harbor, N.Y., 1989, p. B.17.

[5] Hellyer T.J., Brown I.N., Dale J.W., Easmon C.S.F. Plasmid analysis of *Mycobacterium avium-intracellulare* (MAI) isolated in the United Kingdom from patients with and without AIDS. J. Med. Microbiol. 34 (1991) 225–231.

[6] Anonymous. *Gel Filtration: Principles and Methods,* 5th edn., Pharmacia LKB Biotechnology, Uppsala, 1991, pp. 6–13.

[7] Sambrook J., Fritsch E.F., Maniatis T. *Molecular Cloning: A Laboratory Manual,* 2nd edn., Cold Spring Harbor Laboratory, Cold Spring Harbor, N.Y., 1989, pp. E.30-E.38.

Chapter 6

ISOLATION OF LARGE PLASMIDS

Gareth Lloyd-Jones and Peter A. Williams

OUTLINE

Methods in Gene Technology, Volume 2, pages 87–97
Copyright © 1994 JAI Press Ltd
All rights of reproduction in any form reserved.
ISBN: 1-55938-264-3

1. INTRODUCTION

The presence of a plasmid in a bacterial cell may be often inferred by the spontaneous and irreversible loss of a phenotype (such as resistance to an antibiotic or heavy metal or the ability to utilise a particular compound as carbon source) or by the transfer of such a phenotype to a suitable recipient strain by conjugation. Confirmation of plasmid coding requires the physical isolation of plasmid DNA and its correlation with the presence of the particular phenotype. Plasmid isolation is also essential for subsequent molecular studies such as restriction enzyme mapping, hybridisation studies, molecular cloning studies and nucleotide sequencing. Techniques for extracting small plasmids such as are used as vectors for recombinant DNA from *Escherichia coli* and other enteric strains are well developed [1–2]. They can be readily isolated in highly purified form by isopycnic centrifugation in caesium chloride-ethidium bromide gradients, and in a less highly purified form by rapid miniprep methods or by using anion exchange column chromatography cartridges as detailed in Chapters 4 and 5. These isolation procedures rely on the high stability of the supercoiled structure, small size relative to chromosomal DNA, and high copy number of these plasmids.

These same procedures are not suitable for the isolation of large plasmids. This can be accounted for by their low copy numbers, and large size (>100 kb) which makes them inherently more susceptible to single strand nicking and to physical forces such as shear, and also more difficult to separate from linear fragments of chromosomal DNA produced during the preparation. Low copy numbers also necessitate the use of relatively large initial cell masses, which leads to problems associated with separating the plasmid from the large amount of cellular debris formed during the extraction procedure.

Despite these difficulties a large number of plasmids have been isolated, catalogued and characterised from both enteric and non-enteric bacteria, encoding many different phenotypic traits [3], and ranging in size up to 350 kb. Catabolic plasmids, which are generally large (80–300 kb) low copy number plasmids, have been studied from a number of non-enteric bacteria of which *Pseudomonads* appear to predominate [4]. The two methods detailed in this article are successfully used in our laboratory for the isolation, and analysis, of large catabolic plasmids, primarily from *Pseudomonas* strains. The methods provide a means of directly analyzing the restriction profile of large plasmids in excess of 100 kb and yield digested DNA of sufficient purity for (a) analysis of its restriction pattern, (b) hybridisation of Southern blots [5] and (c) use of the digested plasmid for molecular cloning either directly by shotgun cloning or after isolation of specific fragments from agarose gels.

The method of first choice is the Wheatcroft and Williams procedure. This has the advantages of (a) rapidity, allowing digests and cloning to be undertaken the same day that the cultures are harvested and (b) greater yield of DNA, providing sufficient material for digestion, cloning and hybridisations. It does not work for all strains we have studied. It does require plasmids to be large in order that the sucrose gradient centrifugation separates the plasmid from the chromosomal fragments. The smallest plasmid which can be reproducibly isolated is the 60 kb promiscuous IncP1 plasmid known variously as RP4, R68 and RK2 which encodes resistance to kanamycin, tetracycline and penicillins. Plasmids smaller than this separate in about the same fraction as the bulk of the chromosomal fragments. We use the method routinely for plasmids >80–90 kb from *Pseudomonas*. With some bacterial isolates it has proved impossible to obtain adequate plasmid preparations and this may be due to imperfect lysis or the presence of too much macromolecular debris such as polysaccharide. We therefore use the Kado and Liu method as a fall back procedure when Wheatcroft and Williams does not operate, although collaborators of ours have successfully used it as the primary procedure for isolation of catabolic plasmids.

We have had little experience of isolating large plasmids from genera other than *Pseudomonas*, but both procedures should be considered as candidates in situations where the presence of large plasmids is suspected.

2. PLASMID ISOLATION PROCEDURES

2.1 Wheatcroft and Williams Procedure [6]

2.1.1 Basis of Method

Partial denaturation of chromosomal DNA by the combined effects of mild shearing forces and SDS/NaOH treatment followed by separation of the plasmid DNA from the partially denatured chromosomal DNA and cellular debris in a preformed sucrose gradient.

2.1.2   Reagents

Reagent A
   50 mM Tris hydrochloride, without pH adjustment
   50 mM Na$_2$EDTA
   5% (v/v) Dow Corning Antifoam RD emulsion (Hopkin and Williams, Chadwell Heath, Essex, England)
   0.1 mg/mL xylene cyanol (BDH)
   Stable at room temperature for many months.

Reagent B
   1 M NaOH saturated with SDS at 20°C from which the undissolved SDS should ideally be filtered.
   Store at room temperature.
   Make fresh monthly.

Sucrose gradients
   12 mL of 20% w/v sucrose solution, in sterile distilled water, placed in a 14 mL centrifuge tube.
   Gradients are preformed by 2 cycles of freeze/thawing: place gradients in −20°C freezer, remove to room temperature when frozen, allow to thaw completely and freeze again at −20°C. Store frozen until required.

TE buffer
   10 mM Tris-HCl, pH 8.0
   1 mM EDTA

2.1.3   Methodology

1.   Inoculate a single colony from freshly prepared plates of the strain of interest into 2 mL of growth medium. When grown inoculate into 100 mL of either rich medium (nutrient broth, peptone water, Luria broth) or minimal medium with appropriate selection/carbon source as required. This latter may be important in the case of strains the plasmid of which is unstably replicated in the absence of appropriate selection. Grow overnight in an orbital incubator to a final OD$_{600}$ of 0.5–1.0 (>10^7 cells/mL).

2. Harvest the cells by centrifugation, decant the supernatant and resuspend the cell pellet thoroughly in 2 mL of Reagent A by vortexing.

3. Add 0.5 mL of Reagent B. Lyse the cells by gently rotating the tube by hand (approximately 20 rotations/min). Continue for about 1–2 min until the blue colour of the xylene cyanol dye becomes completely green as the cells lyse. The lysed cells will appear viscous due to the presence of intact chromosomal DNA.

4. Chromosomal DNA in the lysed sample is sheared by vigorous vortexing until it is apparent that the viscosity has significantly decreased (approximately 2–5 min): if the solution is still too viscous at this stage, then dilute it with buffer A and vortex again. The surprising feature of this method is that the supercoiled plasmid DNA from large plasmids does not appear to be affected by this step.

5. Layer 2 mL of this preparation onto a preformed sucrose gradient and centrifuge at 100 000*g* in a 16.5-mL swing-out rotor for 1 h at 17°C (decelerating the rotor slowly). After centrifugation, the gradient is fractionated by inserting a glass

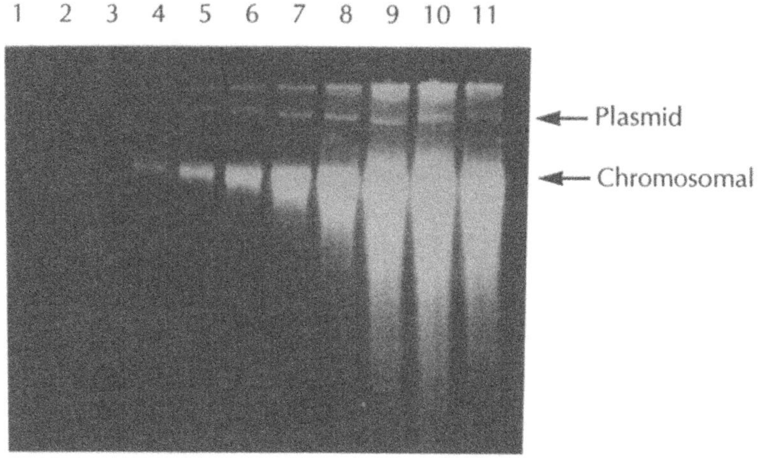

Figure 1. Analysis of fractions from a sucrose gradient. Eleven fractions taken from a sucrose gradient (most dense fractions to the right) and subjected to electrophoresis through a 0.7% agarose gel at 10 V/cm for 6 h. Fractions 6–8 were taken for further analysis. Fractions 9–11 which contain a large amount of chromosomal DNA were usually blue due to the presence of the xylene cyanol dye used in the cell lysis.

tube with attached polythene tubing and syphoning the liquid from the bottom of the gradient, collecting fractions of 1 mL in Eppendorf tubes.

6. An aliquot of each fraction is then electrophoresed in a primary agarose gel to identify the fraction or fractions containing the optimum ratio of plasmid to chromosomal DNA. Figure 1 shows a photograph of a typical gel. It is clear that plasmid DNA is not completely freed from chromosomal contamination, but on subsequent digestion (see Figure 2), the chromosomal DNA is cut into a long smear whereas the plasmid DNA is concentrated into discrete bands.

7. The DNA containing the most concentrated plasmid fractions are further concentrated by centrifugation at 200 000g for 1 h in an angle rotor (by floating the appropriate Eppendorf tubes in

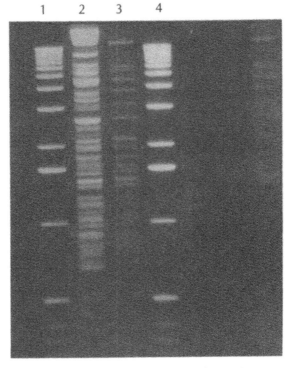

Figure 2.   Restriction analysis of purified plasmids. *Eco*RI digests of a large biphenyl catabolic plasmid pWW100 (track 2) and a deleted derivative pWW100ΔBph (track 3) from a sucrose gradient preparation (as in Figure 1) run alongside a 1 kb ladder (tracks 1 and 4) through a 0.7% agarose gel at 15 mA for 16 h.

water within 10 mL centrifuge tubes).

8. The lower 200 μL is gently removed using a wide bore pipette tip, taking care not to disrupt the gradient, and placed in a fresh Eppendorf tube. The stability of the DNA to storage is very variable and both total DNA and its digests degrade on storage. It is recommended that it normally be used within a few days of preparation and preferably straight away.

9. Plasmid DNA is digested with the appropriate restriction enzyme according to the manufacturers instructions. 10–20 units of restriction endonuclease is usually sufficient. The digests may be electrophoresed immediately but better results are obtained if the digested plasmid is extracted with phenol/chloroform and precipitated with ethanol by conventional means (Chapter 1). The precipitated plasmid digest is finally resuspended in 50 μL TE buffer. The digests are run on agarose gels for 2–3 h or overnight for photography next morning. Figure 2 shows the digests obtained from two large catabolic plasmids prepared by this method.

In our original paper [6] we presented an variant of this procedure which attempted to demonstrate the presence of a plasmid in its native form in the host cell. Although this has been used successfully on occasions (see [8]) we no longer routinely use this as a method, preferring to use the entire procedure above involving the production of reproducible digests to demonstrate definitively the presence of a plasmid. The demonstration of the presence of a large plasmid by visualising it in its native form after agarose gel electrophoresis, may be unsatisfactory for a number of reasons.

(a) Intact plasmid DNA run out on an agarose gel does not always yield a clear picture as the plasmid may be visualised in more than one form (open, nicked or ccc), or confused with the presence of more than one plasmid.

(b) Plasmid DNA may be present at too low a concentration (<2 ng) for the photographic visualisation of the DNA, particularly against the background of a chromosomal smear of DNA.

(c) As a means of size determination the mobility of large uncut plasmids in agarose gels is prone to a large degree of error and it is more accurate to sum the sizes of restriction fragments of the plasmid [8].

The presence of two or more large plasmids within a strain is clearly difficult to distinguish from a digest of the total plasmid DNA but, of course, is also very difficult to distinguish from a gel of undigested plasmid unless the plasmids vary substantially in size. In a number of instances we have used digests to infer such a situation when fragments in the digest disrupt the gradation of the size-intensity relationship expected from a single plasmid (i.e. they are more or less intense than would be expected from their size). This can be attributed to the presence of two or more plasmids which are either present in different copy number or, more likely, are isolated to different degrees of efficiency by the procedure.

2.2 Kado and Liu Procedure [7]

2.2.1 Basis of Method

Total denaturation of chromosomal DNA by a combined heat/SDS/NaOH denaturation procedure with removal of lipid soluble cell debris/SDS and protein by a gentle phenol:chloroform extraction, with the plasmid DNA precipitated in ethanol.

2.2.2 Reagents

Lysis Solution
  2.6 mL of sterile distilled water
  0.16 mL of fresh 2.0 M NaOH
  1 mL of 10% w/v SDS
  0.25 mL 2.0 M Tris base (N.B. not adjusted for pH)
Prepare lysis buffer immediately before use.

TE buffer
  10 mM Tris-Cl, pH 8.0
  1 mM EDTA

RNase
  5 mg/mL in TE buffer

phenol : chloroform (see Chapter 1)

sodium acetate
  3 M sodium acetate, pH 7.5

2.2.3 Methodology

  1. Inoculate 5–10 mL of growth media from fresh colonies (<3

days old) and grow overnight, to an OD_{600} of 0.5–1.0 ($>10^7$–10^8 cells/mL), maintaining the appropriate plasmid selection if required.

2. Harvest 5 mL of the culture by gentle centrifugation and resuspend the cell pellet in 1 mL of sterile distilled water. Transfer to an Eppendorf tube, recentrifuge, discarding the supernatant, and resuspending the cell pellet in a final volume of 75 μl of sterile distilled water.

3. Add an equal volume of fresh lysis solution, mix and incubate at 56–65°C (N.B. Different incubation temperatures seem to suit different plasmids) for 30 min followed by cooling to room temperature (5 min).

4. Add 150 μl of phenol : chloroform and mix the phases gently by inverting the tubes slowly 10 times. Centrifuge for 10 min and recover the upper layer containing the DNA using wide-tipped tips and avoiding any white 'furry' interface consisting of denatured protein and cell debris precipitated by the organic solvent. If the supernatant still appears cloudy, then extract a second time. Load 20–30 μl of the supernatant onto agarose gel and run at 8 V/cm for 1–3 h to visualise plasmid bands.

5. Restriction analysis of the isolated plasmid can now be made. If the plasmid band from step 4 is sufficiently bright, extract any residual traces of phenol from 100–200 μl of sample with water-saturated diethylether until the etherial phase becomes clear. (Use a fume hood for all work with ether.) Evaporate any remaining ether and precipitate the digested DNA with ethanol by conventional means (Chapter 1). Resuspend the dried DNA pellet in TE buffer (20 μl) with 2 μl RNAase and incubate for 30 min at room temperature. Digest with the appropriate restriction enzyme, according to manufacturers instructions.

2.3 Additional Points

2.3.1 *Trouble Shooting*

1. The DNA solution obtained may be too viscous, making it unmanageable and producing smearing on agarose gels. This may be due to:
    (a) Too many cells were used for the preparation.
    (b) The cells used are too old, having reached stationary

   phase, causing build up of polysaccharide.

2. Care must be taken to run samples soon after preparation. Delay may result in DNA degradation. For the Wheatcroft and Williams procedure it is our experience that digested DNA often degrades within a week, although on occasions it may be stable for up to a month at −20°C.

3. If there are problems in the procedures which could be due to the particular strain with which you are working, we recommend that *Pseudomonas putida* mt-2 (also referred to as PaW 1; ATCC 33015; NCIMB 12182; DSM 3931), which contains the 117-kb archetypal TOL plasmid, should be used as a control. Both methods work ideally with this strain and the *Xho*I and *Hin*dIII digests of this plasmid give clear unambiguous gel patterns [9].

2.3.2   Growth media

L-broth

   10 g Bacto-tryptone (Difco)
   5 g Yeast Extract (Difco)
   5 g NaCl
   Make up to 1 L with distilled water, adjust to pH7, sterilise by autoclaving.

Peptone water

   8 g Bacto-peptone (Difco)
   4 g NaCl
   Make up to 1 L with distilled water and sterilise by autoclaving

REFERENCES

[1]   Birnboim, H.C. and Doly, J. A rapid alkaline extraction procedure for screening recombinant plasmid DNA. Nucleic Acids Res., 7 (1979) 1513–1523.

[2]   Maniatis, T., Fritsch, E.F. and Sambrook, J. *Molecular Cloning. A Laboratory Manual.* Cold Spring Harbor Laboratory Press, Cold Spring Harbor, N.Y., 1989.

[3]   Day, M. The biology of plasmids. Sci. Prog., Oxf., 71 (1987) 203–220.

[4]   Sayler, G.S., Hooper, S.W., Layton, A.C. and King, J.M.H. Catabolic plasmids of environmental and ecological significance. Microbial Ecology, 19 (1990) 1–20.

[5]   Southern, E.M. Detection of specific sequences among DNA fragments separated by gel electrophoresis. J. Molec. Biol., 98 (1975) 503–517.

[6]   Wheatcroft, R. and Williams, P.A. Rapid methods for the study of both stable and unstable plasmids in *Pseudomonas*. J. Gen. Microbiol., 124 (1981) 433–437.

[7]   Kado, C.I. and Liu, S.T. Rapid procedure for detection and isolation of large and small plasmids. J. Bacteriol., 145 (1981) 1365–1373.

[8]   Pickup, R.W., Lewis, R.J. and Williams, P.A. *Pseudomonas* sp. MT14, a soil isolate which contains two large catabolic plasmids, one a TOL plasmid and one coding for phenylacetate catabolism and mercury resistance. J. Gen. Microbiol. 129 (1983) 153–158.

[9]   Downing, R.G. and Broda, P. A cleavage map of the TOL plasmid of Pseudomonas putida mt-2. Molec. Gen. Genet. 177 (1980) 189–191.

Chapter 7

ISOLATION OF LINEAR AND SINGLE-STRAND PLASMID DNA

Juan C. Alonso

OUTLINE

Methods in Gene Technology, Volume 2, pages 99–116
Copyright © 1994 JAI Press Ltd
All rights of reproduction in any form reserved.
ISBN: 1-55938-264-3

1.   INTRODUCTION

Bacterial plasmids replicate autonomously and are maintained at a certain average copy number within a population of bacteria under defined growth conditions. Basically two mechanisms by which circular double-stranded (ds) plasmid molecules replicate their own DNA have been described. These modes were conventionally named θ and σ (also termed rolling circle), according to the characteristic structures of replication intermediates. In the θ mode, the leading and lagging strand origins are primed in the near vicinity and both strands remain covalently closed. The elongation step may be uni- or bidirectional and finally the relaxed concatemeric dimer is then resolved into covalently closed monomeric circles (CCC) [1]. In the σ or rolling circle plasmid replication mode a plasmid-encoded product (Rep) makes a single strand (ss) DNA cut at the origin region. The Rep protein remains attached to the 5′ terminus during strand displacement. The 3′-hydroxyl end at the nicking site is extended by DNA PolIII and after one full round of replication, leading strand synthesis terminates by a Rep-mediated cleavage to regenerate the origin. Lagging strand synthesis can be initiated after leading strand synthesis is complete or as soon as the lagging strand replication origin is in a single stranded form, rendering CCC plasmid DNA. The rolling circle replication mode resembles that of single-stranded coliphages [1–4].

   Unit-length single-stranded circles [SS(c)] are replication intermediates of a collection of plasmids from both gram-negative [5–7] and gram-positive bacteria [3,4], which replicate via a rolling circle mechanism. This mode of plasmid replication generates unit-length SS(c) of only one strand (corresponding to the leading strand in replication) [8]. The accumulation of SS(c) DNA is inversely proportional to the efficiency with which lagging strand DNA synthesis initiates in rolling circle replicating plasmids.

   Most of the plasmid molecules, regardless of their replication mode, exist in the monomeric CCC form. Oligomeric plasmid circles accumulate via homologous recombination in certain rec⁻ strains (see [9,10]. In wild type cells bearing a plasmid-borne chi site or in exonuclease V (Exo V) deficient cells, linear concatemeric (LC) plasmid molecules accumulate [11-19]. Linear concatemeric plasmid molecules also accumulate in phage infected cells. The basis for such accumulation is that during late phage replication the dsDNA ends are protected from exonucleolytic degradation either through the attachment of a protein to the dsDNA ends or by interfering with the Exo V activity. Hence, any mechanism which leads to interference of the exonucleolytic activity of the Exo V enzyme leads to the accumulation of LC plasmid DNA. Such LC DNA is composed of linear head-to-tail plasmid molecules which are either ss, ds, ds with ss ends or single-branched circles [13,14,17]. These different types of molecules, with a length ranging from 100 to 650 Kb in size, were observed at a similar frequency [13,14]. From heteroduplex experiments it has been shown that the forks giving rise to the concatemers progress clockwise and/or counterclockwise on the plasmid template, as reported for λrolling circle replication [20–22]. The LC plasmid forms are generated via a recombination-dependent rolling circle replication mode [22].

Recently, some plasmids from a variety of species have been found to be unit-length double-stranded linear DNA molecules. The majority of these have a protein covalently linked to the 5´-end, and replicate in a fashion that resembles that of the *Bacillus subtilis* φ29 bacteriophage (see [23] for a review). The protocols described for the purification of the plasmid LC forms can be used to purify these linear plasmid DNA molecules (see also [24].

Many methods of purification of plasmid CCC forms have been developed and are discussed in detail elsewhere in this volume (Chapters 3–6). This chapter aims at summarizing our knowledge in the purification of other minor plasmid forms such as SS(c) and LC. There are different protocols for the isolation of SS(c) and LC plasmid DNA forms. The methods presented here can be applied to purify SS(c) and LC plasmid DNA from different bacterial species. Many of the techniques have been adapted from published literature. The references cited are either for non-standard techniques, to illustrate the way in which the technique has been used, or to introduce the reader into a technique that is not described here.

2. PRINCIPLE

The SS(c) plasmid form is poorly intercalated by acridine orange, ethidium bromide (EtBr) or propidium diiodide dyes. This is advantageous for purification, but disadvantageous for detection. The buoyant density of the DNA decreases in proportion to the amount of compound intercalated and more dye molecules bind to any chromosomal or plasmid dsDNA form than to the SS(c) form. Therefore, the SS(c) plasmid form is separated from any other double-stranded DNA forms present in a crude lysate by dye-buoyant density centrifugation. Alternatively, the SS(c) DNA can be isolated by hydroxylapatite chromatography. Since SS(c) DNA is usually not detected by fluorescence, Southern hybridization experiments are performed for its analysis.

The isolation procedures normally used to purify the CCC plasmid form which involve DNA denaturation by heat [25] or alkali [26] (see also Chapters 4 and 5) do not yield the LC plasmid form [11,18]. Furthermore, the LC plasmid form co-migrates with chromosomal DNA fragments in the CsCl buoyant density gradient in the presence of intercalating dyes.

Total LC plasmid DNA is usually isolated from Exo V deficient strains. The different DNA plasmid forms are separated by agarose gel electrophoresis and the LC plasmid DNA is then recovered from the agarose gel. Alternatively, a mild lysis method followed by isopycnic centrifugation and rate-zonal sedimentation on sucrose gradients could be used to purify the LC plasmid form. The latter method, however, renders LC plasmid forms of lower molecular weight than the former one.

The protocols presented below are based on established methods for *B. subtilis*. For manipulations with other microorganisms (e.g. anaerobic bacteria) some modifications may be necessary.

3. ISOLATION OF SINGLE-STRANDED CIRCULAR PLASMID DNA

The buoyant density of DNA depends on its G + C content, but the binding of an intercalating dye, such as ethidium bromide, reduces its buoyant density (see Chapter 4). Linear or nicked circular DNA binds more EtBr than CCC DNA. Single-stranded circular DNA binds little if any of the dye. After isopycnic centrifugation the sample in the CsCl gradient is composed of different parts. Two bands of DNA are located in the centre of the tube. The upper band contains the chromosomal DNA and relaxed plasmid forms, whereas the lower band contains the CCC plasmid DNA. The material above the chromosomal DNA band consist of proteins, and the deep red pellet at the bottom of the tube consists of RNA complexes. The SS(c) DNA band, which cannot be directly visualized by ultraviolet light illumination due to the poor intercalation of the EtBr, is placed between the CCC plasmid DNA band and the RNA pellet. In the absence of a precise location the lower part of the gradient (located between the CCC band and the bottom of the tube) is collected. The SS(c) DNA can then be purified further by chromatography.

3.1 Purification of Single-stranded DNA by Caesium Chloride Centrifugation

1. Grow 200 mL of cells bearing plasmid in L-broth (or any other rich media) with a selective drug in a 1 L flask (roughly 5-times the volume of the medium) to about 1.0×10^8 to 2.0×10^8 cells/ml at 37°C.

2. Add rifampicin to a final concentration of 10 µg/ml and incubate further for 30 min. Harvest the cells by centrifugation (10 000g for 10 min at 4°C). Following addition of rifampicin the amount of SS(c) DNA increases 2- to 4-fold when

Table 1.   Buffers used in purification of ssDNA by CsCl centrifugation.

TES buffer 50 mM Tris-HCl pH 8.0 1 mM EDTA 50 mM sucrose	*HS buffer* 50 mM Tris-HCl pH 8.0 1000 mM NaCl 10 mM EDTA
TESN buffer 50 mM Tris-HCl pH 8.0 50 mM sucrose 100 mM NaCl 10 mM EDTA	*PP buffer* potassium phosphate pH 7.5 0.1% Sodium dodecylsulphate
TE buffer 5 mM Tris-HCl, pH 8.0 0.1 mM EDTA	*TED buffer* 50 mM Tris-HCl pH 8.0 200 mM EDTA
TAE buffer 40 mM Tris-Acetate pH 8.0 1 mM EDTA	

compared to the control without the antibiotic [27].

3. Resuspend the pellet in 20 mL of TES buffer (Table 1). Add 2 mL of a 2 mg/mL fresh solution of lysozyme (lysostaphin, mutanolysin or any other equivalent) and 1 mg/mL of RNase A resuspended in TES buffer. Mix the solution gently and incubate for 30 min at 37°C. RNase A at a concentration of 10 mg/mL in 5 mM Tris-HCl pH 8, is made free of DNase by incubation at 100°C for 15 min.

4. Add 2.5 mL of 0.5 M EDTA, pH 8.0. Seal the tube with parafilm and mix the suspension by inverting the tube several times. Keep the tube on ice for 5 min.

5. Add 2 mL of 10% Sarkosyl (N-Lauroylsarcosine). Mix the suspension by inverting the tube several times till the culture clears.

6. Add 0.5 mL of a 5 mg/mL of pronase E. Seal the tube with parafilm, mix the solution by inverting the tube and incubate the tube for 30 min at 37°C. Pronase E is made free of DNases by incubation at 37°C for 60 min.

7. Measure the volume of the DNA solution (remove 50 µl for analysis on a Southern blot). Add 1.1 g of CsCl and 0.1 mL of EtBr (10 mg/mL) for every millilitre of solution.
    CAUTION: EtBr is a powerful mutagen: handle with care.

8. Transfer the solution into a Quick-Seal tube, fill up the tube with paraffin, balance the tubes in pairs and seal them by using the heating device provided by the centrifuge supplier.

9. Centrifuge the tubes for at least 40 h at 250000g, or 18 h at 400000g at 20°C. Two bands of dsDNA, located centre down of the tube, should be visible at ordinary light, but not the SS(c) DNA band. The upper band contains the viscous chromosomal DNA and relaxed plasmid DNA whereas the lower band contains the CCC plasmid DNA. The SS(c) DNA band should be located between the CCC plasmid DNA band and the bottom of the tube.

10. Cut the top of the tube to allow air to enter. Insert a 18-gauge hypodermic needle at the position of the upper DNA band. Collect the viscous DNA into a disposable tube (remove 50 µL). Insert a second needle at the position of the lower DNA band and collect it into a plastic tube (remove 50 µL). Introduce a third hypodermic needle and collect everything that

is between the lower visible DNA band and the bottom of the tube. This fraction should contain the SS(c) plasmid DNA (remove 50 μL for analysis on a Southern blot).

11. Extract the EtBr from the solution by using isopropanol saturated with CsCl solution. Add an equal volume of equilibrated isopropanol and mix the tube content by inversion. Remove and discard the upper pink organic layer, then repeat the extraction until the upper phase is no longer coloured.

12. Dialyse the DNA solution against 10^3-fold excess of TE buffer (Table 1) for 5 to 10 h at 4°C.

13. Concentrate the pooled samples using butyl alcohol. Add an equal volume of butyl alcohol and mix the solution by vortexing. Centrifuge the solution at 3000 g for 1 min. Remove and discard the upper phase. Repeat the step until the desired volume of aqueous phase is achieved.

14. Precipitate the DNA by adding Na-Acetate (pH 4.6) to 300 mM final concentration and two volumes of ethanol pre-cooled to –20°C. Mix by inverting the tube and incubate for 15 min at –70°C. Centrifuge the precipitated DNA at 10 000 g for 10 min at 4°C, then discard the supernatant.

15. Wash the DNA sample in 70 % (v/v) ethanol, centrifuge as in the previous step and dry in vacuo. Dissolve the DNA in TE buffer (Table 1) and store at 4°C. The amount of SS(c) DNA is quantified spectrophotometrically using an extinction coefficient of $8\,780\,M^{-1}\,cm^{-1}$ at 260 nm (OD 1 = 40 μg/mL).

3.2 Purification of Single-stranded DNA by Gel Chromatography

1. Grow the plasmid-containing cells and process them as described in Section 3.1, steps 1 to 6.

2. Dialyse the DNA solution against 10^3-fold excess of 350 mM PP buffer (Table 1) for 5 to 10 h at 4°C.

3. Suspend the gel beads (e.g. HTP-DNA grade, BioRad) at 0.1 g/mL in 350 mM potassium phosphate (pH 7.5), and heat until vigorous boiling, cool down to room temperature and deaerate in a desiccator. The hydrated material will be about 2 to 3 mL per gram of powder. Pour the gel, about 2 ml, into two disposable 0.8 × 4 cm polypropylene columns (e.g. BioRad

poly-prep). Equilibrate one of them with 10 volumes (20 mL/h) of 350 mM PP buffer and the second with 100 mM PP buffer. Do not allow the column to dry.

4. Load the dialysed DNA solution onto the column equilibrated with 350 mM PP buffer (20 mL/h) at room temperature. After the solution has passed through, wash the column bed with 1 volume of the same buffer. The single-stranded molecules are collected in the flow through, whereas the majority of the double-stranded DNA remains bound to the column.

5. Dialyse the flow through from the previous column against a 103-fold excess of 100 mM PP buffer for 5 to 10 h at room temperature. Load the dialysed DNA solution onto the column equilibrated with 100 mM PP buffer. After the solution has passed through wash the column bed with five volumes of 100 mM PP buffer. Join the outlet of the column to a UV detector with wavelength set at 260 nm.

6. Elute the ssDNA by a 100 mM to 350 mM potassium phosphate (pH 7.5) gradient. Monitor the elution of SS(c) plasmid DNA from the column with the UV detector. The SS(c) DNA elutes between 220 mM and 270 mM, whereas the traces of dsDNA usually elute between 350 to 380 mM potassium phosphate buffer pH 7.5. Elution can be performed at low temperature (4°C) provided that the molarity is not higher than 300 mM.

7. Analyse the collected fractions by gel electrophoresis followed by Southern blot hybridization according to published procedures [28]. Pool the fractions that contain the SS(c) plasmid form free of any dsDNA.

8. Concentrate the pooled samples using butyl alcohol. Add an equal volume of butyl alcohol and mix the solution by vortexing. Centrifuge the solution at 3000 g for 1 min. Remove and discard the upper phase. Repeat the step until the desired volume of aqueous phase is achieved.

9. Desalt the concentrated solution though a gel filtration (e.g. LH60 Sephadex) spun column according to published procedures [28].

10. Precipitate the DNA solution as indicated in Section 3.1, steps 14 and 15. A commercially available anion exchange resin which is sold under the trademark of Qiagen (Diagen)

performed well in the purification of the SS(c) plasmid DNA
(see Chapter 5). In this case, follow the manufacturers
instructions.

4   ISOLATION OF CONCATEMERIC LINEAR (LC) PLASMID DNA

The bacterial chromosome ranges in size from 4 000 to 6 000 Kb,
whereas the linear plasmid concatemeric forms range from 100 to
1 000 Kb in length [13,14]. Recently, a very simple technique for
handling large DNA pieces without breaking them was described by
Schwart and Cantor [29]. Upon cell lysis, total LC plasmid DNA is
separated from the bacterial host chromosome by using a
conventional gel electrophoresis. The LC DNA is then recovered
from the agarose gel. If the different subpopulations (ssDNA, dsDNA,
dsDNA with ssDNA ends, single branched molecules, see above) are
to be separated, pulsed-field gel (PFG) electrophoresis should be
used. Conditions for PFG electrophoresis are discussed in detail
elsewhere in this volume (Chapters 11–14).

There are several techniques for the recovery of DNA from agarose,
but only a few are completely satisfactory. The methods chosen here
are: binding of the DNA to powdered glass [30] and electroelution
onto DEAE membrane (see Chapter 3 for alternative procedures).
With these methods one can avoid ethanol precipitation of the DNA.
The dissolving of concatemeric plasmid DNA is time consuming and
shearing can occur.

A method which renders higher quantities of LC plasmid DNA than
the previous one, but with a certain loss of the largest DNA molecules
(due to breakage) is based on a double gradient centrifugation. First,
the LC plasmid DNA is co-purified with chromosomal DNA. Second,
the latter might be digested with different frequent cutter restriction
enzymes which do not cleave the plasmid DNA. Finally, the LC
plasmid DNA is separated from the chromosomal DNA by rate-zonal
sucrose gradient centrifugation.

4.1   Separation of Plasmid DNA Forms

4.1.1   Large Scale

1. Grow 100 mL of plasmid-containing cells in L-broth (or any other rich medium) with a selective drug in a 1 L flask to about 1.0 ×10^8 to 2.0 ×10^8 cells/mL at 37°C. Harvest the cells by centrifugation (10000g for 10 min at 4°C).

2. Resuspend the pellet with 10 mL of TES buffer (Table 1) and centrifuge as in the previous step. Resuspend cells in 3 mL of TES buffer, transfer the suspension into a Petri dish (50 mm diameter) and keep at 4°C.

3. Mix the suspension with 2 mL of a 2% low gelling temperature agarose (e.g. BioRad) and allow to solidify at room temperature. The agarose is prepared in water and kept at 45°C.

4. Cut the solidified gel, which has a thickness of about 3 mm, into 6 to 8 blocks of about 10 mm × 25 mm. Transfer the blocks into a fresh Petri dish.

5. Cover the blocks with lysis buffer (TES containing 0.5 mg/mL lysozyme and 0.1 mg/mL RNase A) and incubate while shaking gently for 2 h at 37°C.

6. Discard the previous solution and cover the agarose blocks with TED buffer (Table 1) containing 3 mg/mL proteinase K and 1% sodium dodecyl sulphate (SDS). Incubate the Petri dish, while shaking gently, for 20 h at 37°C. The DNA released from lysed cells is entrapped in the agarose gel.

7. Discard the previous solution and cover the agarose blocks with TE buffer. Dialyse the DNA entrapped in the small agarose blocks against a 10^3-fold excess of TE for 5 to 10 h at 4°C. If digestion of the DNA is required, transfer the block to a small petri dish (35 mm diameter), cover it with the appropriate restriction enzyme or nuclease buffer and the desired enzyme, and incubate for 2–12 h, at the optimal temperature for the enzyme.

8. Place the agarose blocks into a horizontal gel chamber and pour an 0.6% agarose solution made in TAE buffer (Table 1) over them and allow solidification. One block should contain undigested phage λ DNA embedded with bromophenol blue.

9. Cover the agarose gel with TAE buffer containing ethidium

bromide (200 ng per mL of buffer) to a height of 3 to 4 mm. Separate the different plasmid DNA forms by conventional electrophoresis at 0.5 to 1 V/cm for 20 to 30 h at 4°C.

10. Identify the DNA material by using a long wavelength ultra-violet lamp. The LC DNA should migrate slower than the phage λ DNA.

4.1.2   Small scale

1. Grow 3 mL of plasmid-containing cells in L-broth (or any other rich medium) with a selective drug in a 25 mL flask to about 3.0×10^8 to 4.0×10^8 cells/ml at 37°C. Harvest the cells by centrifugation, using a microcentrifuge at maximal speed for 2 min.

2. Resuspend the pellet in 100 μL of TES buffer containing 1 mg/ml fresh solution of lysozyme (lysostaphin, mutanolysin or any other equivalent) and 20 μg/mL of RNase A and incubate for 60 min at 37°C.

3. Add 100 μL of a 1 mg/mL solution of proteinase K and 1% SDS. Mix the solution by inversion of the tube, and incubate for 60 min at 37°C.

4. Add 200 μL of phenol:chloroform:isoamyl alcohol (25:24:1) to the sample. Mix the contents of the tube until an emulsion forms.

5. Centrifuge the mixture at 6000*g* for 1 min. Transfer the aqueous phase to a dialysis tube.

6. Dialyse the DNA solution against a 10^4-fold excess of TE buffer for 5 to 10 h at 4°C (remove 50 μL for analysis on a Southern blot).

7. Select a restriction enzyme which has no activity against the plasmid DNA, but which digests the chromosomal DNA. Bring the DNA solution to the appropriate buffer conditions and incubate the reaction for 60 min at the temperature optimal for the chosen enzyme (remove 50 μL for analysis on a Southern blot).

8. Separate the different DNA forms by conventional gel electrophoresis in TAE buffer containing ethidium bromide (see above) at 5 V/cm for 2 to 3 h at 4°C.

9. Identify the DNA material by using a hand-held, long wavelength ultraviolet lamp. The LC DNA should migrate slower

than the control, phage λ DNA. Remove 50 μL for analysis on a Southern blot.

4.2   Recovery of Linear DNA from Agarose Gels

4.2.1   Absorption to Glass Powder

1. Cut out the block of the gel containing the high molecular weight plasmid DNA band (LC plasmid DNA) with a scalpel blade and place it in a 1.5 mL microfuge tube.
2. Add 2.5 to 3 times the volume of the solution (or weight of the gel slice) of a 6 M NaI solution in a polypropylene tube. The final NaI concentration has to be close to 4 M. The tube is then incubated for 5 min at 50°C to dissolve the agarose.
3. Add 0.5 mg of powder glass to 1 μg DNA. Mix by inversion of the tube and leave for 5 min at 4°C to enable the DNA to absorb to the glass powder.
4. Spin down the glass powder-DNA complex for 5 s in a microfuge (e.g. in an Eppendorf centrifuge by keeping the button pressed). Avoid overcentrifuging as this leads to the formation of clumps that are difficult to resuspend. Discard the supernatant.
5. Add 0.2 mL of cold (−20°C) wash buffer to the glass powder and resuspend the pellet by gentle vortexing and centrifuge (as in step 4). Repeat the washing cycle twice (remove 50 μL sample).Wash buffer is prepared by mixing equal volumes of absolute ethanol and 20 mM Tris-HCl pH 7.2, 200 mM NaCl, 2 mM EDTA. The presence of ethanol increases the adherence of DNA to glass powder and removes the EtBr.
6. Remove the supernatant with a fine tipped pipette.
7. Resuspend the pelleted glass powder-DNA in 20 μL of TE buffer and incubate at 50°C for 5 min to elute the DNA from the glass powder. Pellet the glass powder as in step 4.
8. Transfer the supernatant solution containing the DNA to a fresh tube (remove 50 μl sample).
9. Add 10 μL of TE buffer to the glass powder pellet and repeat the extraction procedure twice.

Commercially available kits which contain a silica matrix that binds DNA are sold under different trademarks; alternatively, glass powder can be prepared simply and economically in the laboratory. See

Chapter 3 for details of the preparation procedure, and more information on the use of glass powder for recovery of DNA.

4.2.2 Electroelution Onto a DEAE Membrane

1. The positions of the LC DNA and the chromosomal DNA bands are identified as described above (Section 4.1.2,. step 4). Avoid contamination of LC DNA with the chromosomal DNA (which is in the well of the gel) by removing the top of the gel.

2. Make an incision, with a sharp scalpel, in the agarose gel parallel to the lane of interest about 2 mm ahead of it (anode side of the DNA band).

3. Slide in a piece of a wider (2 mm on each side) pre-treated DEAE-membrane (e.g. Schleicher & Schuell NA45) with help of flat forceps (e.g. Millipore forceps) and close the incision by squeezing the front part of the gel against the paper.

   The membrane should be pre-treated by soaking in 10 mM EDTA pH 8.0 for 5 min and then in 0.5N NaOH for 5 min. Wash the membrane in distilled water six to ten times. Petri dishes containing appropriate solutions are suitable for this purpose. Do not allow the pre-treated membrane to dry at any step, otherwise the DNA will bind irreversibly to the dry membrane.

4. Resume electrophoresis (5 V/cm at 4°C) until the band has migrated into the membrane. The migration is followed by using a hand-held, long wavelength ultraviolet lamp.

5. When the desired fluorescent band is trapped on the membrane turn off the power supply. Remove the membrane with flat forceps and rinse the membrane in distilled water to remove any traces of agarose.

6. Place the NA45 membrane into a 1.5 mL microfuge tube. Add 1 mL of isopropanol and invert the tube several times. Remove and discard the supernatant containing traces of ethidium bromide.

7. Add 200 μL of HS elution buffer (Table 1) to cover the membrane and incubate for 30 min at 65°C. The membrane should be folded or slightly crushed to reduce the volume of high salt buffer to be added.

8. Transfer the solution containing the DNA into a fresh tube (remove a sample for analysis). Add a second aliquot of HS

buffer to the membrane, and incubate for 20 min at 65°C.

9. Combine the aliquots of the HS buffer and desalt them through a gel filtration (e.g. LH60 Sephadex) spun column according to published procedures [28].

10 Concentrate the DNA in vacuo (e.g. SpeedVac concentrator, Savant).

4.3 Purification of Linear Plasmid DNA by Centrifugation

4.3.1. Caesium Chloride Isopycnic Centrifugation

1. Grow 200 ml of plasmid-containing cells in mL-broth (or any other rich medium) with a selective drug in a 1-L flask to about 1.0×10^8 to 2.0×10^8 cells/mL at 37°C. Harvest the cells by centrifugation (10 000g for 10 min at 4°C).

2. Resuspend the pellet in 20 mL of TESN buffer containing 1 mg/ml fresh solution of lysozyme (lysostaphin, mutanolysin or any other equivalent) and 20 µg/mL of RNase A and incubate for 30 min at 37°C.

3. Add 2.5 mL of 0.5 M EDTA, pH 8.0. Seal the tube with parafilm and mix the suspension by inverting, very gently, the tube several times. Keep the tube on ice for 5 min.

4. Add 2 mL of TESN buffer containing 500 µg/mL of proteinase K and SDS to 1% final concentration. Mix the suspension by inverting the tube several times until the culture clears and incubate for 30 min at 37°C.

5. Measure the volume of the DNA solution (remove 50 µl for analysis on a Southern blot). Add 1 g of CsCl and 0.1 mL of ethidium bromide (10 mg/mL) per millilitre of solution.

6. Transfer the solution into Quick-Seal tubes, balance the tubes in pairs, if necessary fill the tubes with paraffin and seal them by using the heating device provided by the centrifuge supplier.

7. Centrifuge the tubes for at least 40 h at 226 000g (e.g. 50 Ti rotor or equivalent) or 18 h at 242 000g (e.g. V50 Ti rotor or equivalent) at 20°C. Two bands of dsDNA, located in the centre of the tube, should be visible under ordinary light. The upper band contains linear and open circular DNA whereas the lower band contains the CCC plasmid DNA.

8. Puncture the top of the tube to allow air to enter. Insert a 19-gauge hypodermic needle at the position of the upper DNA

band and collect the DNA material into a plastic tube (remove 50 μL for analysis on a Southern blot).

9. Extract the ethidium bromide from the solution by using isopropanol saturated with CsCl solution. Add an equal volume of equilibrated isopropanol and mix the tube content by inversion. Centrifuge at 3000g for 2 min. Remove and discard the upper pink organic layer, then repeat the extraction until the upper phase is not longer coloured.

10. Dialyse the DNA solution against a 10^4-fold excess of TE buffer for 5 to 10 h at 4°C.

11. Place the dialysed solution into a fresh tube. Digest an aliquot of it with two or three different restriction enzymes which cut the chromosomal DNA very often, but do not cut the plasmid DNA. The restriction enzymes are used as recommended by the suppliers.

12. Extract the sample several times with butyl alcohol to reduce the volume of the sample, desalt the concentrated solution through a gel filtration (e.g. LH60 Sephadex) spun column and concentrate the DNA in vacuo (e.g. SpeedVac concentrator, Savant).

4.3.2 Sucrose Gradient Centrifugation

1. Prepare two or more 38-mL or 12-mL sucrose (10% to 40%) gradients in polypropylene tubes, for Beckman SW28 and or SW41 rotors, respectively. The sucrose solution is made in a buffer containing 20 mM Tris-HCl, pH 8.0, 1 M NaCl, 1 mM EDTA.

2. Load onto each gradient 150 μg of DNA in a volume of about 300 μL when a SW28 rotor is to be used or 50 μg of DNA in a volume of 100 μL in the case of a SW41 rotor.

3. Centrifuge the samples at 112000g for 24 hours in the SW28 rotor or at 210000g for 18 h in the SW41 rotor at 15°C.

4. Collect 0.5-mL fractions through a 18-gauge needle with its tip inserted in the bottom of the centrifugation tube. Aliquots of every second fraction (25 to 50 μL) are mixed with an equal volume of TE buffer and analysed on 0.7% agarose gel. The gels should be run for 2 to 3 h (5 V/cm) and photographed under ultraviolet illumination.

5. Pool fractions containing predominantly linear plasmid DNA and dialyse against water for 5 to 10 h at 4°C.

6. Extract the sample several times with butyl alcohol to reduce the volume and/or concentrate the DNA in vacuo.

5.   TROUBLE SHOOTING

The protocols suggest sampling for either Southern blot analysis when the material is not readily identifiable or direct analysis on an agarose gel when it can be directly detected by UV illumination. This makes possible the localization of the source of any problem that might occur.

5.1   Poor Yield of SS(c) DNA is Obtained

(a) the addition of rifampicin ($10\,\mu g/mL$) and/or the removal from the plasmid dispensable regions, especially the major lagging strand replication origin, lead to a substantial increase in the accumulation of SS(c) DNA.
(b) Test the binding capacity of your hydroxylapatite batch using the coliphage M13 or any equivalent, because its quality is very variable. When higher PP buffer concentrations are required load the column material in a water-jacketed column (e.g. Bio Rad econo-column) and perform elution at 60°C. Prewarm all buffers and solutions.
(c) The removal of phosphate ions improved the efficiency of ethanol precipitation of the DNA.

5.2   The SS(c) DNA is Contaminated with RNA

When removal of RNA traces from the purified SS(c) plasmid DNA is desired, the sample is applied to an anion exchange resin (e.g. Qiagen); RNA and ssDNA elute at about 0.9 M and 1.1 M sodium chloride, respectively.

5.3   Poor Resolution of LC Plasmid DNA

Sample overloading leads to a poor resolution of contaminating DNA from the plasmid linear concatemer. Reduce the amount to be applied into CsCl solution before gradient formation or to be loaded onto preformed sucrose gradients.

ACKNOWLEDGMENT

My sincere thanks to all my students, postdocs and technical assistants who have adapted or developed the techniques described here. This research was partially supported by Deutsche Forschungsgemeinschaft (SFB 344/B5).

REFERENCES

[1] Kornberg, A. and Baker, T. In Kornberg, A. and Baker, T. DNA replication. W.H. Freeman, New York, 1992, pp. 637, 687.

[2] Baas, P.D. DNA replication of single-stranded *Escherichia coli* phages. Biochim. Biophys. Acta 825 (1985) 111–139.

[3] Gruss,A. and Ehrlich S.D. The family of highly interrelated single-stranded deoxyribonucleic acid plasmids. Microbiol. Rev. 53 (1989) 231–241.

[4] Novick, R.P. Staphylococcal plasmids and their replication. Annu. Rev. Microbiol. 43 (1989) 537–565.

[5] Seufert, W., Lurz, R. and Messer, W. A novel replicon occurring naturally in *Escherichia coli* is a phage-plasmid hybrid. EMBO J. 7 (1988) 4005–4010.

[6] Kleanthous, H., Clayton, C.L. and Tabaqchali, S. Characterization of a plasmid from *Helicobacter pylori* encoding a replication protein common to plasmids in gram-positive bacteria. Mol. Microbiol. 5 (1991) 2377–2389.

[7] Yasukawa, H., Hase, T., Sakai, A and Masamune, Y. Rolling-circle replication of the plasmid pKYM isolated from a gram-negative bacterium. Proc. Natl. Acad. Sci. USA 88 (1991) 10282–10286.

[8] teRiele,H., Michel, B. and Ehrlich, S.D. Single-stranded plasmid DNA in *Bacillus subtilis* and *Staphylococcus aureus*. Proc. Natl. Acad. Sci. USA 83 (1986) 2541–2545.

[9] Kolodner, R. Genetic recombination of bacterial plasmid DNA: electron microscopic analysis of in vitro intramolecular recombination. Proc. Natl. Acad. Sci. USA 77 (1980) 4847–4851.

[10] Laban, A. and Cohen, A. Interplasmidic and intraplasmidic recombination in *Escherichia coli* K-12. Mol. Gen. Genet. 184 (1981) 200–207.

[11] Cohen, A. and Clark, A.J. Synthesis of linear plasmid multimers in *Escherichia coli* K-12. J. Bacteriol. 167 (1986) 327–335.

[12] Gruss, A. and Ehrlich, S.D. Insertion of foreign DNA into plasmids from gram-positive bacteria induces formation of high-molecular-weight plasmid multimers. J. Bacteriol. 170 (1988) 1183–1190.

[13] Kusano,K., Nakayama, K. and Nakayama, H. Plasmid med-iated lethality and plasmid multimer formation in an *Escherichia coli recBC sbcBC* mutant. J. Mol. Biol. 209 (1989) 623–634.

[14] Leonhardt, H., Lurz, R. and Alonso, J.C. Physical and biochemical characterization of recombination-dependent synthesis of linear plasmid multimers in *Bacillus subtilis*. Nucleic Acids Res. 19 (1991) 497–503.

[15] Niki, H., Ogura, T. and Hiraga, S. Linear multimer formation of plasmid DNA in *Escherichia coli hopE (recD)* mutants. Mol. Gen. Genet. 224 (1990) 1–9.

[16] Rinken, R. and Wackernagel, W. Inhibition of the recBCD-dependent activa-
      tion of Chi recombinational hot spot in SOS-induced cells of *Escherichia coli*.
      J. Bacteriol. 174 (1992) 1172–1178.
[17] Silberstein, Z. and Cohen, A. Synthesis of linear multimers of *oriC* and
      pBR322 derivatives in *Escherichia coli* K-12: role of recombination and repli-
      cation functions. J. Bacteriol. 169 (1987) 3131–3137.
[18] Viret, J.-F. and Alonso, J.C. Generation of linear multigenome-length plasmid
      molecules in *Bacillus subtilis*. Nucleic Acids Res. 15 (1987) 6349–6367.
[19] Viret, J.-F. and Alonso, J.C. A DNA sequence outside the pUB110 minimal
      replicon is required for normal replication in *Bacillus subtilis*. Nucleic Acids
      Res. 16 (1988) 4389–4406.
[20] Bastia, D., Sueoka, N. and Cox, E. Studies on the late replication of phage λ:
      rolling circle replication of the wild type and a partially suppressed strain,
      Oam29 Pam80. J. Mol. Biol. 98 (1975) 305–320.
[21] Takahashi, S. The starting point and direction of rolling circle replicative inter-
      mediates of coliphage λ DNA. Mol. Gen. Genet. 142 (1975) 137–153.
[22] Viret, J.-F., Bravo, A. and Alonso, J.C. Recombination-dependent concatemer-
      ic plasmid replication. Microbiol. Rev. 55 (1991) 675–683.
[23] Salas, M. Protein-priming of DNA replication. Annu. Rev. Biochem. 60 (1991)
      39–71.
[24] Keen, C.L., Mendelovitz, S., Cohen, G., Aharonowitz, Y., Roy, K.L. Isolation
      and characterization of a linear DNA plasmid from *Streptomyces clavuligerus*.
      Mol. Gen. Genet. 212 (1988) 172–176.
[25] Holmes, D.S. and Quigley, M. A rapid boiling method for the preparation of
      bacterial plasmids. Anal. Biochem. 114 (1981) 193–197.
[26] Birboim, H.C. and Doly, J. A rapid alkaline extraction procedure for screening
      recombinant plasmids. Nucleic Acids Res. 7 (1979) 1513–1523.
[27] Alonso, J.C., Stiege, A.C., Tailor, R.H. and Viret, J.-F. Functional analysis of
      the dna(Ts) mutants of *Bacillus subtilis* plasmid pUB110 as a model. Mol.
      Gen. Genet. 214 (1988) 482–489.
[28] Sambrook, J., Maniatis, T. and Fritsch, E.F. *Molecular Cloning: A Laboratory
      Manual*, 2nd edn., Cold Spring Harbor Laboratory, Cold Spring Harbor, N. Y,
      1989.
[29] Schwart, D. and Cantor, C.R. Separation of yeast chromosome-sized DNAs by
      pulse field gradient gel electrophoreses. Cell 37 (1984) 67–75
[30] Vogelstein, B. and Gillespie, D. Preparative and analytical purification of
      DNA from agarose. Proc. Natl. Acad. Sci. USA 76 (1979) 615–619.

Chapter 8

PREPARATION OF
BACTERIOPHAGE λ DNA

E. Norman

OUTLINE

Methods in Gene Technology, Volume 2, pages 117–125
Copyright © JAI Press Ltd
All rights of reproduction in any form reserved.
ISBN: 1-55838-264-3

1.   INTRODUCTION

Many of the cloning vectors currently in use, particularly for the construction of gene libraries, are based on bacteriophage λ [1,2]. In its wildtype form this is a temperate phage with a head, containing double stranded DNA, and a thin non-contractile tail which requires the presence of the LamB (maltose binding protein) receptor on host bacteria in order to cause infection. The majority of λ based vectors in use have been modified to prevent the integration of recombinant phage DNA into the host chromosome and therefore only exhibit the lytic cycle of infection. In addition the deletion of the t_{R2} λ terminator region (*nin5* or *ninL44*) causes the transcription of the delayed early group of genes, including those required for the packaging and maturation of phage particles and cell lysis, to be independent of the presence of the normally required positive regulator and therefore encourages lysis.

In vectors where the establishment of lysogeny is able to occur (for example in λgt11 and its derivatives which are used for immuno-screening), the *cI* repressor has been modified to be thermosensitive thus allowing for the induction of the lytic cycle by a temperature shift during growth of the lysogen with the subsequent release of phage particles when required. These vectors also carry the amber mutations *Sam7* or *Sam100* in the gene encoding the S protein; this protein is required for lysis of the bacterial host cell. Consequently infection of a non-suppressing host bacterial strain allows the prolonged assembly and accumulation of phage particles which may then be released by lysing the cells with chloroform and therefore enhances the yield of phage DNA or of recombinant fusion proteins obtained. For routine propagation and screening with these vectors the infection of a *supF* host allows the lytic cycle to proceed normally after induction.

There are many methods in use for the preparation of λ phage DNA. This chapter describes two procedures which are adapted from several previously described methods (for examples of other methods see [2,3]) and which should generate DNA suitable for most requirements. One is a method for the large scale preparation of phage DNA from liquid culture and the second is a small scale method using confluent lysis plates as the starting material.

2.   PREPARATION OF BACTERIAL HOST CELLS

The bacterial host strains used for the propagation of phage λ are generally derivatives of *E. coli* strain K12 which are modified to be deficient in the EcoK restriction system. Where specialised vectors are used such as λgt11 and its derivatives special care must be taken to ensure that the host background contains the correct suppressors to permit the required form of phage propagation for example the lytic cycle or delayed lysis. If there is concern over the possibility of deletions or rearrangements occurring within the DNA cloned into a vector then the use of a Rec⁻ host strain may be considered providing that the λ vector in use is *gam⁺* and able to be propagated on this background.

All frequently used host strains may be propagated on standard media (for example, LB) and stock cultures should be well maintained with single colony isolates being used as the inoculum to prepare plating cells to ensure the phage are continually grown on the correct background. As the site of adsorption of λ phage on *E. coli* is the LamB receptor, which is a maltose binding protein, the addition of maltose at 0.2% to the growth media is required to produce host cells suitable for infection with phage. It is also necessary to include a source of magnesium ions in the final host cell preparation as these are essential for phage stability and adsorption. Cells may be prepared in advance by growing an overnight culture at 37°C, centrifuging at 3000 r.p.m. for 10 min to harvest the cells which are then resuspended in 1/2 volume of 0.01 M MgSO₄. When prepared in this way, cells are stable at 4°C for at least one week. Alternatively an overnight culture may be back diluted 1:100, grown to mid-log phase and infected with phage providing that MgSO₄ has been included in the growth medium. In both cases approximately 0.2 mL of cells should be sufficient to generate a lawn on a 9-cm plate. Where a large scale liquid culture is to be inoculated with phage it is also possible to use an overnight culture diluted with fresh medium providing a correct ratio of cells to phage is maintained.

3.   STORAGE AND ASSAY OF PHAGE STOCKS

Small scale phage stocks may be generated by resuspending plaques in SM buffer containing a trace of chloroform; the suspension can be stored at 4°C. These stocks may be amplified by infecting host cells and growing until lysis is seen either in liquid culture or on a confluent lysis plate and then harvesting the released phage and again storing at 4°C in the presence of a trace of chloroform. Phage stocks may be titred to assist in setting up cultures with defined multiplicities of infection by serially diluting the stock, pre-incubating with host plating cells and plating in top agar as described below (Section 4.2).

4.   INFECTION OF THE BACTERIA AND HARVESTING OF THE PHAGE

4.1.   In Liquid Culture

The yield of phage DNA obtained from a liquid culture is related to the extent of lysis which occurs during growth and the success with which the liberated phage particles are harvested.

Cultures may be established with either high or low multiplicities of infection. As recombinant phage may exhibit different rates of growth the optimal multiplicity of infection and time of growth may need to be adjusted for different phages. Cultures with a high multiplicity of infection in the region of 1 : 5 (phage:cells) may achieve complete lysis within 2–3h. Those established with a low multiplicity of infection, for example between 1 and 10 phage per 200 cells will require several additional rounds of infection and lysis before the majority of host cells have been infected and lysed. Therefore the time taken to reach complete lysis is longer and the risk of loss of phage due to adsorption to cellular debris is increased. Once lysis has occurred cell debris must be removed and the phage particles present in the supernatant be recovered by PEG precipitation [4]. The following protocol describes the growth and harvesting of phage from a high multiplicity of infection culture but with adapted phage to bacteria ratios and growth times this would be suitable for use with low multiplicity of infection cultures.

1.   Grow a suitable sensitive bacterial strain overnight in LB medium supplemented with maltose and $MgSO_4$.

2.   To 20 mL of bacterial culture, add 10^9 phage in λ diluent or SM

buffer and allow the phage to adsorb at 37°C for 15 min.

3. Use this to inoculate 200 mL of the same medium and grow at 37°C, with shaking at 250 r.p.m. for 2–4 hours, checking periodically for lysis.

4. Once lysis has occurred add 10 mL of CHCl₃ and continue to shake at 37°C for 15 min to ensure that all bacteria have lysed.

5. Centrifuge at 9 000 r.p.m. for 15 min at 4°C to pellet cellular debris.

6. Recover the supernatant and add 20 g PEG 6000 and 5 g NaCl, dissolve at room temperature and then leave to stand at 0°C for a minimum of 4 h.

7. Centrifuge at 10 000 r.p.m. for 30 mins at 4°C to pellet the precipitated phage particles.

8. Resuspend the phage particles in 4 mL of λ diluent, add 0.75 g CsCl per mL, dissolve and centrifuge in a Beckman Ti50 rotor or equivalent at 38 000 r.p.m. for 24 h at 4°C. A bluish band of phage particles should be visible after centrifugation. Alternatively the phage may be purified on CsCl or glycerol step gradients [2].

9. Recover the phage band from the gradient, dilute to a final volume of 3 mL per gradient with λ diluent and proceed to extract and purify the phage DNA.

4.2. On Solid Medium

Again, as with growth in liquid media, most standard *E. coli* media may be used providing they are supplemented with both maltose and MgSO4. However as agar contains compounds which may persist through and contaminate the final DNA preparations and which may be inhibitory to restriction enzymes agarose is substituted for agar.

1. Prepare plating cells as described above.

2. To 0.2 mL of plating cells, add approximately 10^5 phage in λ diluent or SM buffer and incubate at 37°C for 15 min.

3. Add this to 3 mL of Top LB agarose at 45°C, mix and immediately pour onto the surface of a pre-dried agarose LB plate. Allow the overlay to set, invert the plate and incubate it at 37°C for 5–6 hours until confluent lysis is seen. Overnight incubation of plates can be detrimental and may result in the

growth of bacterial colonies on the surface of the plates and hence to a high level of bacterial debris being present in the phage suspension after harvest.

4. Flood the plate with 5 mL of λ diluent plus approximately 0.2 mL CHCl₃ and leave at room temperature with gentle agitation for 1–2 h. Prolonged incubation at this stage appears harmful and may result in later difficulty in digesting the DNA with restriction enzymes.

5. Recover the phage suspended in λ diluent with a pasteur pipette and centrifuge at 9 000 r.p.m. for 10 mins at 4°C to remove cellular debris. Once removed from the plate the phage suspension may be stored overnight at 4°C if required.

6. Recover the supernatant and add RNase and DNase I to final concentrations of 1 μg mL⁻¹ each. Incubate at 37°C for 30 min to digest any DNA or RNA of bacterial origin present.

7. Add an equal volume of a solution containing 20% PEG 6000 and 2 M NaCl in λ diluent and stand in wet ice for a minimum of 1 h

8. Centrifuge at 10 000 r.p.m. for 20 min at 4°C to recover the phage particles.

9. Discard the supernatant taking particular care to remove all traces of PEG. Failure to remove all PEG at this stage may result in difficulty in cutting the resulting DNA with restriction enzymes. Resuspend the phage pellet in 0.5 mL of λ diluent and proceed to extract and purify the phage DNA.

5. EXTRACTION AND PURIFICATION OF PHAGE DNA.

1. Distribute the phage particles in λ diluent in 0.5 mL aliquots in microcentrifuge tubes and add 5 μl of 10% SDS and 5 μl of 0.5 M EDTA pH 8.0. Incubate at 68°C for 15 min.

2. Add an equal volume of phenol: CHCl³ : isoamylalcohol (25:24 : 1), vortex well and centrifuge at room temperature in a microcentrifuge for 1 min or until the phases are well separated.

3. Recover the aqueous phase and repeat the extraction with phenol: CHCl₃ :isoamylalcohol as above.

4. Recover the aqueous phase and extract with an equal volume

of CHCl₃: isoamylalcohol (24 : 1) as above.

5. Add two volumes of absolute ethanol to precipitate the DNA and centrifuge at room temperature for 15 min.
6. Resuspend the pellet in an appropriate volume of TE buffer and add 1/2 volume of 7.5 M ammonium acetate plus 2 volumes of absolute ethanol to reprecipitate the DNA. Centrifuge at room temperature for 20 min.
7. Wash the pellet with 70% ethanol, dry and resuspend in TE buffer and store at 4°C or −20°C until use.

N.B. At this stage the phage DNA preparation contains a significant amount of RNA and therefore RNase should be included in restriction digests. Alternatively an RNase digestion on the whole sample may be carried out followed by a further cycle of extraction with phenol: CHCl₃:isoamylalcohol and ethanol precipitation.

6. YIELD

Different λ phage vectors and different recombinant clones can exhibit markedly different growth rates following infection into susceptible cells. Where difficulty is encountered in producing lysis either on plates or in liquid culture then a reassessment of the titre of the phage stock in use to check that cultures are being established with a correct multiplicity of infection for the growth period being allowed may be helpful.

The genotype of the bacterial strain being used as a host should be also be verified to ensure that all necessary host mutations are present to permit the growth of the phage vector in question. Particular care should be taken to check that the host background contains the correct suppressor mutations. As the yield which will be obtained from λ phage DNA preparations varies greatly depending upon the degree to which lysis occurs it is difficult to give reliable estimates. Figures of between 500 µg to several milligrams of DNA per litre of culture have been quoted for standard methods [2]. For DNA preparations from confluent lysis plates by the method described here the yield is normally in the region of 5 µg DNA per harvested plate.

7.   MEDIA AND SOLUTIONS

LB Medium
   Tryptone, 10 g
   Yeast extract, 5 g
   NaCl, 10 g

Make up to 1 L and adjust to pH 7.5 with NaOH.

For solid medium, add 15 g agarose.
For top agar, add 8 g agarose.
After autoclaving add 10 mL of 1 M $MgSO_4$ plus 10 mL of 20% maltose.

Maltose
   Stock solution made up at 20% and filter sterilized.

SM buffer
   NaCl, 5.8 g
   $MgSO_4.7H_2O$, 2.0 g
   1 M Tris.Cl (pH 7.5), 50 mL
   2% gelatin solution, 5 mL

   Make up to 1 L and sterilize by autoclaving.

λ diluent
   $MgSO_4.7H_2O$, 2.0 g
   1 M Tris.Cl (pH 7.5), 10 mL
   0.5 M EDTA (pH 8.0), 0.2 mL

   Make up to 1 L and sterilize by autoclaving.

TE buffer
   10 mM Tris.Cl (pH 7.6)
   1 mM EDTA (pH 8.0)

SDS
   10% (w/v) solution in water

EDTA
   0.5 M EDTA, adjusted to pH 8.0 with NaOH.
   Sterilize by autoclaving.

Phenol/chloroform/isoamylalcohol
   phenol/chloroform/isoamylalcohol (25 : 24 : 1 by volume)

Chloroform/isoamylalcohol
chloroform/isoamylalcohol (24 : 1 by volume)

Ammonium acetate
7.5 M ammonium acetate
Sterilized by autoclaving

REFERENCES

[1] Kaiser, K. and Murray, N.E. The use of phage λ replacement vectors in the construction of representative genomic DNA libraries. In Glover, D.M. (ed.), *DNA Cloning. II : A Practical Approach*, IRL Press Oxford, 1985, pp. 1–47

[2] Sambrook, J., Fritsch, E.F. and Maniatis, T. *Molecular Cloning. A Laboratory Manual*, 2nd. edn., Cold Spring Harbor Laboratory Press, Cold Spring Harbor, N.Y., 1989.

[3] Arber W., Enquist, L., Hohn, B., Murray, N.E. and Murray, K. Experimental methods for use with lambda. In Hendrix, R.W., Roberts, J.W., Stahl, F.W. and Weisberg, R.A. (eds.), *Lambda II*, Cold Spring Harbor Laboratory, Cold Spring Harbor, N.Y. 1983, pp. 436–466

[4] Yamamoto, K. M. and Alberts, B.M. Rapid bacteriophage sedimentation in the presence of polyethylene glycol and its application to large scale virus purification. Virology, 40 (1970) 734–744.

Chapter 9

PURIFICATION OF VIRAL RNA

J.E. Whitby, A.D.T. Barrett, A.D. Jennings
and P. Johnstone

OUTLINE

Methods in Gene Technology, Volume 2, pages 127–149
Copyright © JAI Press Ltd
All rights of reproduction in any form reserved.
ISBN: 1-55938-264-3

1. INTRODUCTION

The purity and integrity of RNA isolated from virions is fundamentally important to its successful use in subsequent molecular procedures. Such procedures include cDNA cloning, *in vitro* transcription, reverse transcription and amplification by the polymerase chain reaction (RT-PCR), primer extension sequencing and Northern blot analyses.

The relative purity of extracted viral RNA (vRNA) varies depending on the source of virions used, the degree of interaction between capsid proteins and RNA, and on the extraction protocol employed. vRNA extracted from gradient purified virions can result in a very pure sample of vRNA, whereas total RNA extracted from virus infected cell culture, or from a tissue sample, will result in a mixed population of cellular and vRNA. The degree of contamination of the vRNA sample by protein, lipid and deoxyribonucleotides will vary depending on the stringency of procedures used for its extraction. Furthermore, the level of RNAases present in different cell and tissue samples will affect the ultimate success rate of obtaining undegraded vRNA.

Techniques such as primer extension sequencing and cloning both require highly purified undegraded vRNA, whereas less purified samples of RNA are often adequate for RT-PCR. Clearly the origin of the viral sample and the use to which the extracted vRNA will subsequently be put will dictate the choice of extraction procedure.

All protocols for the successful deproteinisation and isolation of vRNA follow a number of important steps:
- purification of the virions
- disruption of the virions
- denaturation of the nucleoprotein complexes

- inhibition of the ribonucleases
- purification of the vRNA from virion debris

However all the protocols fall into three main categories, those that use guanidinium salts [1], those that use phenol [2], and those that use sodium dodecyl sulphate (SDS) and proteinase K [3]. There are a substantial number of variations, permutations and combinations of these three basic categories.

In this chapter, a number of tried and tested procedures for the extraction of vRNA from virions, using all the categories of methods, will be described. However, a number of variations on standard procedures will not be described, as excellent descriptions exist elsewhere. Where appropriate, references to other suitable techniques will be supplied. Furthermore, a number of kits have been developed which are reliable and convenient and preclude the need to use time-consuming and hazardous older methodologies. These will also be described briefly.

2. PURIFICATION OF VIRIONS

The first step in vRNA extraction is the appropriate purification and enrichment of the virus. The methodologies for this vary widely, depending on the size and morphology of the virus and the degree of purification and enrichment required. Therefore the new researcher is well advised to refer to scientific publications in his/her field to ascertain the most common method of purification relevant to a particular virus. However, a few general guidance notes will be included here which are of general interest.

Appropriate containment precautions must be taken at all stages with viruses known to be pathogenic. To separate virions from the cellular debris of host cells, low speed refrigerated centrifugation is performed ($10\,000g$ for 10–$20\,min$ at $4°C$).

Aggregation and concentration of virions may be effected by two main methods, ammonium sulphate or polyethylene glycol (PEG) precipitation. Ammonium sulphate is added to a final concentration of 40% (w/v) to the infected tissue culture fluid. Pelleting is achieved by low speed centrifugation and the pellet is resuspended in at least 10% of the original volume to ensure adequate dilution of the ammonium sulphate. This method is rapid, achieves a tenfold concentration of the virus and has a recovery of greater than 99%. PEG is added to a final concentration of 2.5–10% (w/v), depending on the morphology of the virus at 4°C and precipitation is enhanced by the addition of 0.1 M sodium chloride. Pelleting is then effected by low-speed centrifugation. This method achieves greater concentration of virus and avoids the fluctuations in pH associated with ammonium sulphate precipitation. It is, however, more time consuming. Alternatively, pelleting of the virus may also be effected by high speed centrifugation alone, for times and at speeds dependent on the size and morphology of the virus. This final method involves no addition of chemicals and achieves a purer preparation of virus but it is only suitable for small volumes (less than 200 mL) of virus.

Further purification of virions is most usually effected by gradient centrifugation on rate zonal and/or isopycnic gradients. In rate zonal centrifugation the position of the virus within the gradient is governed by its sedimentation coefficient and in isopycnic centrifugation, by its buoyant density. A variety of gradients may be used including sucrose rate zonal gradients, caesium chloride and caesium sulphate isopycnic gradients and potassium-tartrate and glycerol rate-zonal/isopycnic gradients. Alternatively, iodinated compounds such as metrizamide or Nycodenz (Nygaard and Co.) can be used for centrifugation [4]. These compounds can achieve excellent separation; however it should be remembered that nucleic acids have lower buoyant densities than proteins in iodinated compounds and the position of the virus band in such gradients may be altered. More details of the theory and practice of gradient centrifugation may be found in Hull [5].

However, it is possible to purify and enrich total cellular RNA and vRNA directly from cell culture by extraction with guanidinium salts. Furthermore this is usually the method of choice when the source of the virions is a tissue or an organ sample. If the cells used for growing virions are low in RNAases, it is possible to use other quicker methods. Several such methods are described including a methodology for the extraction of total cellular RNA and vRNA from cell culture grown virions using SDS and proteinase K.

3.   INHIBITION OF RIBONUCLEASES

The successful extraction of RNA from any origin is extremely dependent on the elimination of ribonuclease (RNAase) activity. RNAases are particularly stable enzymes which are resistant to boiling, are active over a wide range of pH and have no cofactor requirements [6]. They are small enzymes composed of a single polypeptide chain [7] and are present in almost all cells, to differing degrees. Their ability to renature easily after many denaturation protocols [8], coupled with their stability, makes them ubiquitous.

Consequently, any work with RNA must assume their presence and stringent steps must be taken to prevent their contact with the nucleic acid. They may be introduced into the RNA sample from naked skin, air-borne dust, glassware, plasticware, pipettes, untreated chemicals and solutions. Furthermore, the cells in which viruses are cultured will all contain RNAases, to a greater or lesser extent, and disruption of these cells will result in their release from membrane bound organelles.

At all times, when working with RNA, disposable gloves must be worn and care must be taken not to introduce contamination by contact with other surfaces. Similarly it is important to dedicate an appropriate set of laboratory equipment for use in RNA work only.

A number of procedures have been developed to remove RNAases from laboratory containers and solutions and these must be adopted if successful extraction of intact RNA is to be achieved.

3.1 Treatment of Glassware

All glassware should be purchased new and reserved for RNA work only. Furthermore it must be treated with a solution of diethyl pyrocarbonate (DEPC) which is a strong, but not absolute, inhibitor of RNAases [9]. DEPC is an irritant to the skin, mucous membranes and respiratory tract and all manipulations must be contained within a fume hood.

Glassware (and lids) should be treated by soaking for a minimum of twelve hours in a 0.1% (v/v) solution of DEPC in distilled water. The solution is then poured off, the containers are rinsed with DEPC treated distilled water and autoclaved for 15 min at 15 lb/sq inch on the liquid cycle. We routinely repeat this autoclaving step to be sure of the complete removal of all traces of DEPC.

Autoclaving inactivates the DEPC to alcohol and water, thus removing traces of DEPC which would otherwise modify purine residues of the RNA by carboxymethylation.

Alternatively glassware may be baked at 180°C for a minimum of 8h before use.

3.2   Treatment of Plasticware

This should be bought gamma irradiated, and therefore essentially RNAase free, wherever possible. Where this is not feasible, the plasticware must be treated with DEPC as described.

All tips and microcentrifuge (MCC) tubes should be purchased new, handled only with disposable gloves and placed into RNAase-free containers. It is useful to remember that plasticware in sealed gamma-irradiated bags is sterile and normally 'untouched by human hand' until opened. Thus for many uses it may be convenient to use such plasticware directly from the bag, without further handling.

3.3   Treatment of Chemicals and Solutions

It is possible to treat solutions for inactivation of RNAases. The solution is made 0.1% with DEPC and complete removal of RNAases is effected by stirring or vigorous shaking for 15min. The solution is then autoclaved twice to inactivate the DEPC.

There are two drawbacks to DEPC treatment of solutions. Firstly, DEPC reacts with adenine residues of RNA rendering it inactive in *in vitro* translation reactions [10] and therefore great care must be taken to remove all traces from solutions. Secondly, DEPC also reacts with Tris and so all Tris buffers cannot be DEPC-treated.

We observe the following points when making up solutions. All chemicals are reserved for RNA work only. Spatulas used for weighing are baked at 250°C for 3h before use and chemicals are measured directly into a DEPC treated container. We routinely use high performance liquid chromatography (HPLC) grade water (Aldrich) for the preparation of all our solutions. We have found this system to be trouble-free and no further treatment of solutions is performed.

3.4 Inhibitors of RNAases

Once the vRNA has been extracted and other RNAase inhibitors, such as guanidinium and phenol have been removed, any RNAases present are able to renature. RNAase activity can be controlled by the maintenance of the sample on ice and by the addition of RNAase inhibitors. Two types of RNAase inhibitors are currently in widespread use, RNAasin and vanadyl-ribonucleoside complexes (VRCs).

We routinely use RNAasin [11] which is an enzyme isolated from either rat liver or human placenta, and forms a RNAase-RNAasin complex. This inhibitor is used at concentrations of 250–2000units/mL and requires the presence of sulphydryl reagents for maximum activity [12]; 1 mM dithiothreitol (DTT) is routinely added. RNAasin is removed by phenol and is compatible with a wide range of subsequent uses of the RNA, including reverse transcription and cell-free translation studies.

VRCs [13] are also extremely effective inhibitors of RNAases. They are added to intact cells at a final concentration of 10 mM which is subsequently maintained throughout the RNA extraction. They are more difficult to remove, but this may be effected by multiple extractions with phenol containing 0.1% 8-hydroxyquinoline Furthermore, EDTA and SDS cause the RNAse-VRC to dissociate and thus this form of inhibition is not reliable for extraction methods which employ these reagents.

4. VIRAL RNA EXTRACTION PROTOCOLS

4.1 Proteinase K-Sodium Dodecyl Sulphate-Phenol Extraction

Sodium dodecyl sulphate (SDS) is a powerful ionic detergent which lyses cells and releases nucleic acids from the normal associations with protein and lipid found *in vivo*. The anionic detergent also exerts a mild inhibitory effect on endogenous RNAases and furthermore its negative charge prevents its interaction with RNA. However, the use of SDS requires the use of higher temperatures (10°C) in order to prevent the detergent from precipitating out of solution. This presents obvious problems for extraction of vRNA from tissues containing high levels of RNAases which are best carried out on ice in the presence of more powerful RNAase inhibitors.

Other non-ionic detergents, including Nonidet P-40 (NP-40), Triton X-113 and Triton X-100, may be used at a concentration of 1%. These milder detergents are generally employed for unenveloped viruses which are however, most usually associated with host membranes when taken from a tissue culture sample.

Following SDS treatment, further dissociation of protein from vRNA may be achieved by incubation with proteinase K [3]. Finally, the isolation of vRNA, relatively free of contaminating oligodeoxyribonucleotides, is effected by extraction with phenol and phenol:chloroform pH 5.1. [2]. The use of the low pH ensures removal of the DNA into the organic phase and interphase, leaving the vRNA in the aqueous phase. It should be noted that phenol extractions (not phenol:chloroform extractions) of polyadenylated RNA at pH values below 7.6 will partition the RNA into the organic phase. However, the presence of SDS ensures good recovery of polyadenylated RNA into the aqueous phase regardless of pH values. This combination of extractions reduces losses of vRNA to the interface produced by the formation of insoluble RNA-protein aggregates [14]. Furthermore, the low pH employed during these extractions, reduces the level of activity of any RNAases present.

The success of this simple and less time consuming methodology is dependent on the performance of all steps as quickly as possible.

Clearly the best results will be obtained when this methodology uses virions which have been purified away from the cellular debris and endogenous RNAases (most usually effected by gradient centrifugation). However, a successful outcome for this procedure, which allows for successful RT-PCR of the resultant vRNA, has been achieved by the use of a crude purification of virions. The method for this is given in Section 4.1.2.

4.1.1   Buffers and Reagents

TNE buffer
   100 mM Tris-HCl pH 8.5
   50 mM sodium chloride
   1 mM disodium EDTA

TE buffer
   10 mM Tris-HCl pH 8
   1 mM disodium EDTA

20 mM Dithiothreitol (DTT)
   Dissolve 0.062 g DTT in 20 mL of 0.01 M sodium acetate pH 5.2.

Sterilise by filtration through a 0.2 µM filter and store in small aliquots at –20°C.

Proteinase K

Dissolve 0.1 g Proteinase K in 5 mL of HPLC grade water. Store in aliquots at –20°C.

10% SDS solution

Dissolve 10 g of SDS in 80 mL of HPLC grade water (SDS should be weighed out in a fume hood). Heat to 68°C to dissolve. Adjust the final volume to 100 mL. (Before use this solution should be warmed at 50°C in a hot water bath if a white precipitate is visible).

ACE buffer

10 mM sodium acetate
50 mM sodium chloride
3 mM disodium EDTA pH 5.1

ACE saturated phenol : 8-hydroxyquinoline

Care must be taken when handling phenol. Phenol burns can be very serious and operators should take appropriate precautions including wearing goggles and gloves. As a further precaution, it is advisable to place the phenol container in a tray to contain any spillages on the bench.

Saturate molecular biology quality phenol with 0.3 volumes of ACE buffer and mix well. Add 0.1% 8-hydroxyquinoline and store in aliquots at –20°C. Immediately before use, the phenol should be thawed at 60–65°C.

Ace saturated phenol:chloroform:isoamyl alcohol:8-hydroxyquinoline (P:C:I) (25:24:1:0.1)

ACE buffer equilibrated phenol is mixed with an equal part of chloroform : isoamyl alcohol (24 : 1) and mixed well. This solution should be stored in foil-covered glass bottles at 4°C.

70% ethanol

Add 70 mL of ethanol to 30 mL of HPLC grade water. Store at –20°C.

5M sodium chloride

Add 29.22 g of sodium chloride to 100 mL of HPLC grade water. Dissolve and autoclave.

Ether

Ether is highly volatile and extremely flammable and must be handled only in an explosion-proof hood. It is preferable to look for safer alternatives

Add HPLC grade water to 25 mL of ether in a 100 mL bottle, mixing well, until a layer of water forms at the bottom of the bottle. Store in the fume hood.

4.1.2   Crude Purification of Virus

This methodology was developed for the purification of flaviviruses from infected cell culture. Some adjustment to the centrifugal time and force may therefore be necessary for other viruses.

1. Clarify the infected cell culture suspension by centrifugation at 10 000g for 35 min at 4°C (a suitable rotor is the Beckman JA14).
2. Decant the supernatant into fresh tubes and sediment the virus by ultracentrifugation at 141 000g for 6 h at 4°C (SW28 rotor).
3. Drain the pellet of supernatant and add 100 μL TE pH 8. Seal the tube with Nescofilm, and allow the pellet to swell at 4°C overnight. Vigorously vortex to resuspend the virions prior to use or storage at −70°C.

4.1.3   RNA Purification

1. Dispense the purified virions into 50 μL aliquots in MCC tubes on ice and make the total volume up to 500 μL with ice-cold TNE buffer.
2. Add 5 μL of 20 mg/mL proteinase K and incubate for 30 min at 37°C.
3. Add 51 μL of 10% SDS, mix and incubate for 15 min at 37°C. Do not place the sample on ice as this will cause the SDS to precipitate out.
4. Add 900 μL of ACE saturated phenol : 8-hydroxyquinoline. Shake vigorously by hand for 5 min. Centrifuge at 5 000g for 5 min and remove the aqueous phase to a fresh tube, taking great care to avoid the interphase.
5. Back extract the organic phase by the addition of 200 μL of TE pH 8. Again mix thoroughly by hand and centrifuge as described. Remove the second aqueous phase and combine it with the first.

6. Add an equal volume of phenol:chloroform:isoamyl alcohol:8-hydroxyquinoline (P:C:I) to the pooled aqueous phases. Mix thoroughly by hand, centrifuge and remove the aqueous phase as described above. If necessary re-extract the aqueous phase with P:C:I until no white flocculent interphase is present (usually only one further P:C:I extraction is required).

7. In a fume hood, add an equal volume of ether (saturated with HPLC grade water), mix by two inversions and then centrifuge at 5 000g for 1 min at room temperature. Remove and discard the upper ether phase as completely as possible and then boil off any residual ether by incubating the tube, lid open, at 68°C in a waterbath for 5–10 min.

8. Add 2.5 volumes of chilled ethanol (–20°C) and 0.04 volumes of 5 M sodium chloride to the RNA suspension and mix thoroughly by inversion. Precipitate the RNA at –70°C for 30 min or in a dry ice/methanol bath for 10 min. Pellet the RNA by centrifugation at 12 000g for 20 min at 4°C. Remove the supernatant and wash the pellet in 200 µL of 70% chilled ethanol and then centrifuge for 1 min at 12 000g at 4°C. Thoroughly remove the supernatant and dry the pellet very briefly in a cool oven (40°C) or in a vacuum desiccator.

9. Resuspend the pellet in 20 µL of HPLC grade water by gentle vortexing. Add 40 U RNAasin and 1 µL of 20 mM DTT. Quantitate the RNA as described in Section 6 and store in appropriate aliquots at –70°C.

Notes:

(i) It is important to keep the volume of the aqueous phase bulked up during the P : C : I extractions and this is performed by adding appropriate measured quantities of TE pH8 as necessary. This makes it easier to avoid the interphase whilst removing the aqueous layer and minimises loss of RNA.

(ii) It is important not to dry the RNA pellet (Step 7) too much as dried pellets of nucleic acid are difficult to redissolve. If difficulty is experienced, then a brief heating at 65°C may expedite dissolution.

(iii) Where the predicted quantity of RNA is very low, 5 µg of carrier tRNA may be added prior to the organic solvent extractions (Step 4) to enhance RNA recovery. However, this makes it difficult to quantify the yield of purified vRNA.

(iv) Ethanol precipitation of RNA from aqueous solutions may be performed as described, at –20°C overnight, at –70°C for 30 min or in a dry-ice/methanol bath for 10 min. The method of choice is often dictated by convenience. However, when dilute samples of RNA are present, overnight precipitation at –20°C is usually very effective. The choice of salt is determined by the nature of the sample and by its subsequent use. Most routinely, a 0.1 volume of sodium acetate pH 5.2 (0.3 M) is added. However, where SDS is present in the sample, sodium chloride (0.2 M) is used as described above. Under these conditions SDS remains soluble and is removed with the 70% ethanol. Other salts and solvents may be used depending on requirements and for more details the reader is referred to Sambrook *et al.* [15].

4.2   SDS and Phenol Extraction

Phenol extraction followed by ethanol precipitation is a common and popular method of vRNA extraction [1]. It is faster and much simpler than other methods, but the yield of RNA is less predictable.

The inclusion of SDS in this basic methodology has been found to enhance dissociation of the proteins from the vRNA.

This method is essentially very similar to that given in Section 4.1.3 with only the first three steps differing. On completion of the two steps below, the methodology follows that given in Section 4.1.3. beginning with Step 4.

4.2.1   Buffers and Reagents

See Section 4.1.1.

4.2.2   RNA Purification

1.   The purified virions are dispensed into 10–50 µL aliquots (not exceeding a protein concentration of greater than 5 mg of viral protein) and the total volume is made up to 500 µL with TNE buffer.
2.   SDS is added to the virus suspension to a final concentration of 1% and the suspension is heated to 50°C for 5 min.

4.3   Hot Phenol Extraction for the Isolation of RNA from Nuclei

Hot phenol extraction [16,17] produces good quality RNA which is relatively free of contaminating DNA. This is particularly useful when the vRNA is extracted from preparations of nuclei where DNA is a significant contaminant. However this methodology does not produce vRNA sufficiently pure for a number of purposes, including cloning and primer extension sequencing.

This method is suitable for virions which are found preferentially in nuclei or for viral mRNA (from DNA or RNA viruses) found in the nucleus.

4.3.1 Buffers and Reagents

(See also Section 4.1.1.)

Washing buffer:
   10 mM Tris pH 7.9
   10 mM sodium chloride
   5 mM magnesium chloride
   100 mM potassium chloride

Extraction buffer
   1% SDS
   100 mM Tris-HCl pH 7.4
   20 mM disodium EDTA.

NETS buffer
   10 mM Tris-HCl pH 7.4
   100 mM sodium chloride
   10 mM disodium EDTA
   0.2% SDS

High salt buffer
   10 mM Tris pH 7.4
   0.5 M sodium chloride
   50 mM magnesium chloride
   2 mM calcium chloride

Phosphate buffered saline
   8 g sodium chloride
   0.2 g potassium chloride
   0.24 g potassium dihydrogen phosphate
   Dissolve in 800 mL distilled water, adjust the pH to 7.4 with hydrochloric acid and add distilled water to 1 L. Autoclave and store at room temperature.

NP-40 Lysis Buffer
   10 mM Tris pH 7.9
   140 mM potassium chloride

5 mM magnesium chloride
1 mM DTT
0.2-0.5% (v/v) NP-40

4.3.2   RNA Purification

(a) Production and resuspension of infected cell monolayer:
The virus is grown in monolayers of the appropriate cell lines in tissue culture flasks. The quantities described below are for a single 175-cm^2 flask and thus volumes should be adjusted appropriately for flasks of a different size.
1. Pour off the tissue culture fluid and wash the cell sheet with 20 mL of ice-cold PBS. Remove the cell layer by trypsinisation or mechanically, by the addition of 2 mL of ice cold PBS and scraping with a rubber policeman. (If large quantities of extracellular virus are produced, then the supernatant may be saved for later RNA extraction).
2. Add 8 mL ice-cold PBS to the cells, mix well, and pellet by centrifugation at 2000g for 5 min at 4°C.
3. Repeat Step 2 to wash the cells.

(b) Isolation of nuclei:
1. Place the cells on ice and gently resuspend in NP-40 lysis buffer to a final concentration of 5×10^7 cells/mL. Maintain on ice for 5 min. The final concentration of NP-40 is normally between 0.2 and 0.5% but this must be determined for individual cell types using phase-contrast microscopy to confirm lysis of the cytoplasmic membranes.
2. Pellet the nuclei by centrifugation at 1000g for 3 min at 4°C.
3. Remove the supernatant and wash the nuclei with washing buffer containing DNAase at 50 units/mL (see Step 5) and pellet again as for Step 2. Place on ice.
4. Resuspend the nuclei at a concentration of 5×10^7 nuclei/ml in high salt buffer to effect lysis. The viscosity of the solution will increase markedly upon liberation of nucleic acid.
5. Draw the suspension up into a 18 gauge hypodermic syringe needle to shear liberated chromosomal DNA. Repeat (three or four times) until the viscosity of the suspension is reduced by adequate shearing. The concentration of DNAase may need to be adjusted to ensure this is effected within 30 s in order to

limit degradation of the RNA.
6.   Add an equal volume of extraction buffer.

(c) Hot phenol extraction
1.   Add an equal volume of phenol : 8-hydroxyquinoline saturated with NETS buffer followed by an equal volume of chloroform:isoamyl alcohol. The contents of the tube are vortexed briefly and then the tube is heated to 65°C in a waterbath for 10 min with frequent thorough shaking. Chill the tube on ice.
2.   Centrifuge at 1000g for 5 min at 4°C. Remove and discard the bottom organic phase (leaving the interphase and aqueous phase in the tube) and repeat Steps 1 and 2.
3.   Add 2 volumes of chloroform:isoamyl alcohol, mix by inversion and then centrifuge at 1000g for 5 min at 4°C. Remove the aqueous phase to a clean tube.
4.   Precipitate the RNA by the addition of 2.5 volumes of ethanol (–20°C) and then pellet at 12 000g for 20 min at 4°C. Carefully remove the supernatant and wash the pellet in 200 μL of 70% chilled ethanol. Centrifuge at 12 000g for 1 min at 4°C. Thoroughly remove the supernatant and dry the pellet very briefly in a cool oven (40°C) or in a vacuum desiccator.
5.   Resuspend the pellet in 20 μL of HPLC grade water by gentle vortexing. Add 40 U RNAsin and 1 μL of 20 mM DTT. Quantitate the RNA as described in Section 6 and store in aliquots at –70°C.

Note:
(i)   The purpose of the shearing in Step 5 (Isolation of the nuclei) is to reduce the DNA in size and not to totally degrade it. This is done in order to reduce the viscosity of the solution. The subsequent hot phenol extraction will remove the DNA fragments.

4.4   SDS and Phenol Exxtraction of Cytoplasmic RNA from Infected Cell Monolayers

A useful protocol for the extraction of RNA from the cytoplasm of cells is described in the paper by Preston [18] and the reader is directed to this paper for the methodology. VRCs and high concentrations of Tris are used to inhibit RNAases. The separation of nuclei and cytoplasm is effected by pipetting and centrifugation followed by SDS and phenol chloroform extraction of RNA. See also

Sanders *et al.* [19] for further discussion of this and other methods for extraction of cytoplasmic and polyadenylated RNA.

4.5   Guanidinium Thiocyanate Extraction from Monolayers of Infected Cells

The extraction of vRNA from cells containing high levels of RNAases presents obvious problems. Guanidinium salts and β mercaptoethanol (βME) together produce one of the most potent inhibitory systems of RNAases known [8]. Cellular proteins, including RNAases, readily dissolve in concentrated solutions of guanidinium chloride (>4M) losing all secondary structure [1] and β-ME further acts to disrupt the intramolecular protein bonds [8]. Furthermore, guanidinium salts act on the viral nucleoprotein complexes, forcing relaxation of protein secondary structure and releasing vRNA into solution free of protein. This combination of reagents is therefore employed in a number of methodologies for the extraction of all types of RNA [1,20,21].

The method described is adapted from that of Chirgwin *et al.*, [20] and is based on the differential precipitation of RNA from DNA and other cellular components based on their relative solubility in different guanidinium salts.

This procedure is not the method of choice for dilute solutions of RNA.

4.5.1   Buffers and Reagents

(See also Section 4.3.1.)

Guanidinium Thiocyanate Stock Solution
  4 M guanidinium thiocyanate
  0.5% sodium N-lauroylsarcosine
  25 mM sodium citrate pH 7
  0.1 M β-mercaptoethanol

Adjust to pH 7 with sodium hydroxide. Filter through a Whatman No. 1 filter paper. Store tightly sealed at room temperature for up to 1 month. All handling of this solution must be performed in a fume hood.

Guanidine Hydrochloride Solution
  7.5 M guanidine hydrochloride, neutralised to pH 7 and buffered by

the addition of 0.025 volumes of 1 M sodium citrate pH 7. Filter through a Whatman No. 1 filter paper. Add DTT to a final concentration of 5 mM. Store at room temperature for up to 1 month.

3M Sodium Acetate pH 5.2

Dissolve 40.81 g of sodium acetate.3 H_2O in 80 mL HPLC grade water. Adjust the pH to 5.2 with glacial acetic acid. Adjust to 100mL with HPLC grade water. Sterilise by autoclaving.

4.5.2   RNA purification

1. Monolayers of infected cells should be processed as given in Section 4.3.2, steps 1–3.
2. Resuspend the cell pellet, by immediate and thorough vortexing in 750 µL of 4 M guanidinium thiocyanate solution. Add a further 750 µL of 4 M guanidinium thiocyanate solution to the resuspended cells and again vortex thoroughly. At this stage the cells from a number of flasks may be pooled and the appropriate volume additions scaled up accordingly.
3. Draw the suspension up into a 18 gauge hypodermic syringe needle to shear liberated chromosomal DNA. Repeat (three or four times) until the viscosity of the suspension is reduced by adequate shearing.
4. Centrifuge at 10 000*g* at 10°C for 10 min and transfer the supernatant to a new tube.
5. Add 37.5 µL of 1 M acetic acid to lower the pH from 7 to 5 and 0.5 volumes of cold ethanol (–20°C). Mix thoroughly by shaking and then place at –20°C overnight to precipitate the RNA.
6. Centrifuge at 10000*g* for 10min at 4°C to pellet the RNA.
7. Thoroughly remove the supernatant and immediately add 750µL of guanidine hydrochloride solution. Shake the sample vigorously to thoroughly resuspend the pellet. (Brief heating at 68°C in a waterbath may be necessary to effect complete dissolution of the pellet after this first precipitation). The RNA is not free of RNAases at this stage and rapid redissolution of the pellet in guanidine hydrochloride solution is very important to prevent RNA degradation.
8. Reprecipitate the RNA by adding 37.5 µL of 1 M acetic acid

and 0.5 volumes of cold ethanol (–20°C). Place the tube at –20°C for a minimum of 3 h and centrifuge the tube at 10 000g for 10 min at 4°C to pellet the RNA.

9. A further re-precipitation is performed in the same way (steps 6-8), but the total volume of guanidine hydrochloride added is halved and the RNA is pelleted at 10 000g for 5 min only.

10. Remove the supernatant and thoroughly resuspend the pellet in 250 μL of ethanol at room temperature. Centrifuge for 10 min at 10 000g to pellet the RNA. Thoroughly remove the ethanol and dry the pellet briefly in a cool oven.

11. Resuspend the RNA pellet, by vigorous shaking, in 75 μL of 10 mM disodium EDTA pH 7.

12. Add 225 mL chloroform:1-butanol (4 : 1 v : v) and vortex thoroughly. Centrifuge briefly to separate the phases.

13. Remove the upper aqueous phase to a fresh tube on ice. Re-extract the organic phase plus interface (the interface will contain most of the RNA) by adding 1 mL of 10 mM disodium EDTA pH 7, vortexing, centrifuging briefly to separate the phases, and adding the aqueous phase to the earlier one. These re-extractions should be repeated until the size of the interphase layer no longer decreases (generally three in total). Pool the aqueous phases.

14. Precipitate the RNA by the addition of 0.1 volumes of 3 M sodium acetate pH 7 and 2.5 volumes of ethanol (–20°C). Precipitate the RNA at –20°C overnight. Centrifuge for 20 min at 12 00g at 4°C. Carefully remove the supernatant and wash the pellet in 200 μL of 70% chilled ethanol. Centrifuge for 1 min at 12 000g at 4°C. Thoroughly remove the supernatant and dry the pellet very briefly in a cool oven (40°C) or in a vacuum desiccator.

15. Resuspend the pellet in 20 μL of HPLC grade water by gentle vortexing. Add 40 U RNAsin and 1 μL of 20 mM DTT. Quantitate the RNA as described in Section 6 and store in aliquots at –70°C.

Notes:
(i) If the presence of residual oligodeoxyribonucleotides in the vRNA sample is not a problem for subsequent procedures, then the final pellet from Step 11 may be resuspended as given in Step 15. However, if Steps 12–14 are not performed then the RNA sample is

likely to contain contaminating oligodeoxyribonucleotides which may subsequently anneal to RNA during cloning or primer extension sequencing to produce erroneous priming events.

(ii)   Further purification of polyadenylated RNA from total cell RNA may be effected, also without the removal of residual oligodeoxyribonucleotides as described above. This is performed using chromatography on oligo-dT-cellulose [22,23] and an excellent description of this is given in Sambrook [15]. We have not, however, found this to be a particularly efficient methodology. A commercial kit exploiting magnetic separation using biotinylated oligo-(dT) primers and streptavidin paramagnetic particles is routinely used in this laboratory (See Section 5).

(iii)   The key to a successful outcome for this protocol lies in an efficient inactivation of the RNAases before they can degrade the RNA. Thus immediate and thorough mixing upon the addition of the guanidinium thiocyanate, which lyses the cells, is very important.

(iv)   It is important to remember that in the presence of guanidinium salts RNAase activity is inhibited. However, its removal is accompanied by renaturation and a resumption of activity by any remaining RNAases and appropriate steps must be taken to minimise this. These include maintaining the sample on ice and the addition of RNAase inhibitors.

(v)   If problems are encountered at Step 5 with guanidinium thiocyanate crystallising out at –20°C, then 0.75 volumes of ethanol should be added. This, however, does increase the co-precipitation of some protein and DNA.

4.6   Other Protocols using Guanidinium Salt Extraction of RNA

Extraction of vRNA from infected tissue or organs using guanidinium thiocyanate is described in the paper by Chirgwin *et al.* [21]. The method is essentially similar to that described above except that the starting point is a tissue or organ sample. Due to the importance of inactivating the RNAases released from the lysed cells before they can degrade the RNA, vigorous homogenization of a fresh sample in guanidinium thiocyanate solution, is required at the beginning of the procedure.

Mention should also be made of a popular methodology in which cells are disrupted using guanidinium thiocyanate followed by a separation of RNA from the guanidinium homogenate by ultracentrifugation through a caesium chloride cushion [20,24,25]. The buoyant density of RNA is much higher in caesium chloride than that of DNA and other cellular components. The RNA is therefore pelleted to the bottom of the tube and isolated from other components. This methodology is excellent for use on very dilute samples of RNA or for very scarce valuable samples. Excellent descriptions of this methodology are given in Sambrook [15] and in Chirgwin *et al.* [20].

Finally isolation of total RNA from eggs has not been discussed. The reader is directed to Sambrook [15] for a methodology which uses Proteinase K, SDS and phenol extraction followed by lithium chloride precipitation.

4.7   RNA Extraction using Kits

A number of kits are now on the market which facilitate the extraction of RNA with minimum of effort. Such kits save much time in DEPC treatment of glassware and reagents and produce good quality RNA. Most kits available are based on extraction using guanidinium salts. Such kits include the RNAzol and RNAzol B kit (Biogenesis Ltd.) and Promega's RNAgents total RNA isolation kit. Both these kits have extraction protocols based on a modification of the guanidinium thiocyanate methodology and they both isolate total RNA from a wide variety of sources, including tissue and cells.

5.   ISOLATION OF POLYADENYLATED RNA FROM TOTAL RNA

In cases where the vRNA is polyadenylated, its subsequent isolation may be effected by the use of affinity chromatography on oligo (dT)-cellulose [22,23]. Clearly this methodology will separate out both the polyadenylated vRNA and the 1–5% mRNA which is usually extracted in total cell RNA.

Such purification is preferable when preparing vRNA as templates for cDNA libraries. Better results will also be obtained in hybridisation studies.

A methodology for the preparation of oligo(dT)-cellulose is described by Gilham [26] and the protocol for its use is described in Sambrook [15]. However we routinely obtain superior results using a commercially prepared biotinylated-oligo (dT) probe and streptavidin paramagnetic particles (Promega) with subsequent separation by a magnetic separation stand. This system has been used with a good degree of success in this laboratory.

6. ANALYSIS OF PURIFIED RNA

vRNA may be quantitated in a number of ways:

(i)   By the spectophotometric method [27]. This method is routinely employed for nucleic acid samples greater than 500 ng and provides an estimate of both quantity and quality. The principle disadvantage of this method is its failure to distinguish between degraded and undegraded nucleic acid. The absorption of the vRNA sample at 230, 260 and 280 should be determined. An optical density of 1 at 260 nm is equivalent to 40 ng/mL for RNA. A ratio of 230/260 of 2.3 and 280/260 of 2 indicates pure preparations of RNA have been derived.

(ii)   The vRNA may be directly visualised by electrophoresis in a gel containing formaldehyde [15]. However, this procedure requires a large quantity of RNA to be visualised. Furthermore, formaldehyde denatures the secondary structure of the RNA which may affect the subsequent usage of any recovered RNA.

(iii)   RT-PCR may be used directly to check the integrity of the RNA, a successful RT-PCR reaction confirming the integrity without the need for electrophoresis. This method allied to spectophotometric analyses is the method preferred by the authors in cases where the vRNA is valuable and the viruses grow to a low titre.

7. STORAGE OF RNA

Purified vRNA may be stored resuspended in HPLC grade water containing 40 U RNAsin and 1 mM DTT. Alternatively 3 volumes of ethanol may be added to the resuspended pellet. In either case the samples are stored at −70°C in appropriate aliquots in order to avoid frequent thawing and refreezing of the samples.

REFERENCES

[1]   Cox, R.A. The use of guanidinium chloride in the isolation of nucleic acids. Methods Enzymol., 12B (1968) 120–129.

[2]   Stallcup, M.R. and Washington, L.D. Region-specific initiation of mouse mammary tumour virus RNA synthesis by endogenous RNA polymerase II in preparations of cell nuclei. J.Biol. Chem., 258 (1983) 2802–2807.

[3] Hilz, H., Wiegers, U. and Adamietz, P. Stimulation of proteinase K action by denaturing agents: application to the isolation of nucleic acids and the degradation of masked proteins. Eur. J. Biochem., 56 (1975) 103–108.

[4] Barrett, A.D.T., Crouch, C.F. and Dimmock, N.J. Defective interfering Semliki Forest virus populations are biologically and physically heterogeneous. J. Gen. Virol., 65 (1984) 1273–1283.

[5] Hull, R. Purification, biophysical and biochemical characterisation of viruses with especial reference to plant viruses. In *Virology, a Practical Approach*, IRL Press Oxford, 1985, pp. 1–24.

[6] Davidson, J.N. *Biochemistry of the Nucleic Acids*, 8th edn., Chapman and Hall, London, 1976, pp. 163–200.

[7] Blackburn, P. and Moore, S. Pancreatic ribonuclease. In *The Enzymes*, Vol XV. Academic Press, London, 1982, pp. 317–433.

[8] Sela, M., Anfinsen, C.B. and Harrington, W.F. The correlation of ribonuclease activity with specific aspects of tertiary structure. Biochim. Biophys. Acta., 26 (1957) 502–512.

[9] Fedorcsak, I. and Ehrenberg, L. Effects of diethyl pyrocarbonate and methyl methanesulfonate on nucleic acids and nucleases. Acta Chemica Scandinavica, 20 (1966) 107–112.

[10] Ehrenberg, L., Fedorcsak, I. and Solymosy, F. Diethyl pyrocarbonate in nucleic acid research. Prog. Nucleic Acid Res. Mol Biol., 16 (1976) 189–262.

[11] Blackburn, P., Wilson, G. and Moore, S. Ribonuclease inhibitor from human placenta. Purification and properties. J. Biol. Chem., 252 (1977) 5904–5910.

[12] de Martynoff, G., Pays E. and Vassart, G. The synthesis of a full length DNA complementary to thyroglobulin 33S messenger RNA. Biochem. Biophys. Res. Commun., 93 (1980) 645–653.

[13] Berger, S.L. and Berkenmeier, C.S. Inhibition of intractable nucleases with ribonucleoside-vanadyl complexes: Isolation of messenger ribonucleic acid from resting lymphocytes. Biochemistry 18 (1979) 5143–5149.

[14] Palmiter, R.D. Magnesium precipitation of ribonucleoprotein complexes. Expedient techniques for the isolation of undegraded polysomes and messenger ribonucleic acid. Biochemistry 13 (1974) 3606–3615.

[15] Sambrook, J., Fritsch, E.F. and Maniatis, T. *Molecular Cloning: A Laboratory Manual*. 2nd edn, Cold Spring Harbor Laboratory Press, Cold Spring Harbor, N.Y., 1989.

[16] Soeiro, R. and Darnell, J.E. Competition hybridization by pre-saturation of HeLa cell DNA. J. Mol. Biol., 44 (1969) 551–562.

[17] Nevins, J.R. Isolation and analysis of nuclear RNA. In *Guide to Molecular Cloning Techniques*, Academic Press, London, 1987, pp. 234–241.

[18] Preston, C.M. The cell-free synthesis of herpesvirus-induced proteins. Virology, 78 (1977) 349–353.

[19] Sanders, P.G., Jackson, T., Newcombe, J., Bell, S., Scopes, G.E. and Knight, A. Extraction and purification of eukaryotic mRNA. In Grange, J.M., Fox, A. and Morgan, N.L. (eds.), *Genetic Manipulation: Techniques and Applications*,Technical Series No. 28, Society for Applied Bacteriology, 1991, pp. 113–127.

[20] Chirgwin, J.M., Przybyla, A.E. Macdonald, R.J. and Rutter, W.J. Isolation of biologically active ribonucleic acid from sources enriched in ribonuclease. Biochemistry, 18 (1979) 5294–5299.

[21] Macdonald, R.J. Swift, G.H. Przybyla, A.E. and Chirgwin, J.M. Isolation of RNA using guanidinium salts. Methods Enzymol., 152 (1987) 219–227.

[22] Edmonds, M., Vaughan, M.H. and Nakazato, H. Polyadenylic acid sequences in heterogeneous nuclear RNA and rapidly-labeled polyribosomal RNA of HeLa cells: Possible evidence for a precursor relationship. Proc. Natl. Acad. Sci., 68 (1971) 1336–1340.

[23] Aviv, H. and Leder, P. Purification of biologically active globin messenger RNA by chromatography on oligothymidylic acid-cellulose. Proc. Natl. Acad. Sci., 69 (1972) 1408–1412.

[24] Glisin, V., Crkvenjakov, R. and Byus, C. Ribonucleic acid isolated by caesium chloride centrifugation. Biochemistry, 13 (1974) 2633–2637.

[25] Ullrich A., Shine, J. Chirgwin, J., Pictet, R., Tischer E., Rutter, W.J. and Goodman, H.M. Rat insulin genes: Construction of plasmids containing the coding sequences. Science, 196 (1977) 1313–1319.

[26] Gilham, P.T. Synthesis of polynucleotide celluloses and their use in fractionation of polynucleotides. J. Am. Chem Soc., 86 (1964) 4962–4985.

[27] Maniatis, T., Fritsch, E.F. and Sambrook, J. *Molecular Cloning: A Laboratory Manual*. Cold Spring Harbor Laboratory Press, Cold Spring Harbor, N.Y., U.S.A., 1982.

Chapter 10

LARGE SCALE DNA SEQUENCING BY MANUAL METHODS

Andrew J. Davison and Elizabeth A.R. Telford

Methods in Gene Technology, Volume 2, pages 151–175
Copyright © 1994 JAI Press Ltd
All rights of reproduction in any form reserved.
ISBN: 1-55938-264-3

1. INTRODUCTION

The sizes of today's largest DNA sequencing projects are far beyond those imagined when the bacteriophage M13-dideoxynucleotide technology was first developed by Sanger [1] and Messing [2]. Several sequences well in excess of 100 kb have been published, and the largest at present, that of yeast chromosome III, extends to 350 kb [3]. The labour intensive nature of large scale sequencing projects, even those employing the most modern approaches, is apparent from the 'cast of thousands' author lists associated with the majority of resulting publications. The desire to increase throughput and reduce labour in order to obtain very large sequences (chromosomes and even entire eukaryotic genomes) has been much discussed in the popular scientific press, and has stimulated automation of existing technology and promoted investigation of more radical approaches that have not yet reached practical application.

Commercial sequencing machines have been a useful addition to the armoury, and regular upgrading of their capacity is encouraging. At present, however, they are expensive, quite labour intensive and not markedly more efficient than manual techniques. Our aim is to describe how large scale sequencing projects may be undertaken manually (in 'hand-to-hand' combat, as it were) with minimal costs in consumable items. We set out our particular approach, which has been successful for targets in the 130–150 kb range, and have not attempted to describe other methodologies with which we are not experienced. Although now somewhat diverged, our methods are derived from the excellent protocols given in references [4] and [5], which we consider to be essential reading. They should not be regarded as static, however, since it is our practice to streamline them continually. We emphasize that although this chapter is geared to those who wish to undertake large scale projects, the principles and methods described can be adapted readily to more modest targets.

2. PRINCIPLES

2.1 The Sequencing Method

Random fragments of target DNA are cloned into a bacteriophage M13 vector and sequenced by specific chain termination. Target DNA is broken into random fragments, and fragments larger than a predetermined size are selected, repaired enzymatically to ensure that they have blunt ends and ligated into an M13 cloning vector at a blunt-ended site within the *lacZ* gene. Ligated DNA is transfected into competent male *Escherichia coli* and plated onto bacterial lawns, and recombinant plaques are detected by their inability to synthesize β-galactosidase. Small quantities of single-stranded bacteriophage DNA ('templates') are prepared, and each is subjected to dideoxynucleotide sequencing.

This involves annealing a complementary oligonucleotide primer just outside the inserted DNA, and extending it using a DNA polymerase. The reactions are carried out in the presence of four deoxynucleoside triphosphates [2′-deoxy-7-deazaguanosine 5′-triphosphate (7-deaza-dGTP; a dGTP analogue which reduces hydrogen bonding and eliminates many secondary structure artifacts), dTTP, dCTP and radiolabelled dATP] and one of the four 2′,3′-dideoxynucleoside 5′-triphosphates (ddNTPs); four separate reactions are required for each template, each containing a different ddNTP as chain terminator. The ddNTP:dNTP ratio ensures that specific termination of the extending DNA strand occurs in a proportion of DNA molecules, resulting in a nested set of radiolabelled products sharing a common 5′end. These are resolved by electrophoresis on high resolution denaturing gels and visualized by autoradiography. The sequence data are 'read' into a computer using a digitizer, and are then compiled into a database by a computer-driven process of overlapping individual sequences. Reading and compilation errors are edited from the database by referring to the original autoradiographs, and remaining difficulties are resolved by specific experiments. Finally, the completed DNA sequence is analysed for its coding potential.

2.2   Efficiency

In determining a large DNA sequence as quickly as possible, it is important to establish the optimal balance between throughput of templates and the quality of individual gel readings. Emphasis on one factor tends to be detrimental to the other and, if excessive, will thwart success. The methods described below place emphasis on throughput, and individual workers should set the rate of processing templates so that a satisfactory quality is maintained.

The sole aim of a sequencing project should be to obtain a correct sequence. Thus, the only exciting point comes at the end, and the temptation to analyse the coding potential of part or all of an incomplete database can become very strong. Unless there is a good scientific reason, it is best to resist. The time spent carrying out partial analyses can be excessive, and delays completion of the project. For greatest efficiency, it is advisable to treat the various stages of a sequencing project as parts of a batch process, and to analyse the data only at the very end.

2.3   Cost Effectiveness

Only a modest amount of equipment is required in addition to that normally available in research laboratories. Moreover, the methods described below have a low cost in consumables. The chemicals required are inexpensive, with the exception of commercial sequencing primers, the large proteolytic subunit of E. coli DNA polymerase I ('Klenow enzyme') and radioisotope. Costs associated with the first two items can be reduced significantly. Modern oligonucleotide synthesizers are able to produce relatively huge amounts of primers of sufficient quality for direct use in sequencing without further purification. Our current 1 μmol batch of universal primer cost about £50 to make and will suffice for about 200 000 templates (i.e. in excess of the number required to sequence a bacterial genome). Also, large amounts of sequencing grade Klenow enzyme can be produced with ease from a genetically engineered strain of *E. coli* [6]; a single batch can last for tens of thousands of templates. Radioisotope now forms by far the major part of our consumable expenses, but even this can be minimized by reducing the amount added to each sequencing reaction and increasing the exposure times of autoradiographs; a cost of £20 of isotope per finished kb of sequence is feasible.

These considerations are not intended to give an underestimate of the expense of DNA sequencing. It is true that the price of materials can be reduced to a level that is insignificant compared with the cost of labour. Nevertheless, labour costs are real, and that is why the most important ingredient in a sequencing project is a worker who has an able pair of hands and is able to focus in a determined way on the final goal.

2.4 Accuracy

The data from autoradiographs are read into a computer, where the database is assembled. Since reading autoradiographs is inherently a fallible process, the database accumulates errors. Therefore, there must be stages at which all errors are identified and removed. In this respect, two approaches to reading data may be taken. In the first, ambiguity codes are used when the identity of a particular nucleotide is in doubt, so that each gel reading is as accurate as possible. Greater effort is needed in reading gels with such care, but a more accurate database is obtained. Also, errors are more easily identified in the database, and editing requires only occasional reference to the autoradiographs. The second approach, which we follow, relies on the fact that a database can accommodate a significant proportion of errors and still not cause problems to data entry. Less emphasis is placed on accuracy, ambiguity codes are not used and 'best guesses' are made in regions of difficulty. Thus, the rate of reading gels is maximized, but much more care is needed in correcting errors at the editing stage. We check every nucleotide, on both strands where possible, by referring to at least one autoradiograph.

Random sequencing gives information from the majority, but usually not the entirety, of both DNA strands. Data from both strands should be obtained for as much of the sequence as possible, since sequence from only one strand may contain problems caused by secondary structure effects. A problem of this sort is usually obvious and can be resolved by specifically obtaining data from the other strand. On rare occasions, it may go undetected and lead to incorporation of an error into the completed sequence. For this reason, most journals pragmatically require sequences to be determined fully on both strands. This is not onerous for smaller sequences, but the time taken and the expense incurred to complete both strands for a larger sequence can be prohibitive. Since errors may arise from several additional sources, and thus no sequence, however carefully determined, can be considered error-free, it is our view that personal judgment of data quality (particularly in single-stranded regions) should be the key factor in deciding how far to proceed in sequencing both strands fully

3. APPROACH

3.1 M13-dideoxynucleotide Sequencing

DNA for sequencing may come from a variety of sources, but most studies involve preliminary subcloning of a region of interest into a plasmid, cosmid or bacteriophage lambda vector. Targets in this size range (up to 40 kb) constitute small to moderate scale projects. Larger sequences may be determined by linking together several smaller sequences. Methods for isolating recombinant DNA in its various forms and for treating DNA with restriction endonucleases are not within our remit, and readers should refer to one of the many standard laboratory texts.

Often, the target insert DNA is isolated from the vector after treatment with restriction endonucleases, self-ligated to remove ends and sonicated [4]. For cosmids, it can be more efficient to sonicate circular DNA and sequence the insert and vector together. In some circumstances, large DNA molecules may be isolated in pure form and sequenced without the need for intermediate subcloning. This has several advantages, not the least of which is increased efficiency [7].

We are involved in sequencing herpesvirus genomes with sizes in the range of 125–200 kb by this route [8, 9; E.A.R. Telford, unpublished data]. It is likely that larger molecules of up to a few million bp lacking extensively reiterated regions (e.g. prokaryotic genomes) could also be achieved by directly subcloning random fragments into M13 vectors.

The amount of work required to complete a project is approximately proportional to the purity of the starting DNA. Therefore, careful effort should be made to obtain DNA of an appropriate quality. Also, since large DNA sequencing projects involve processing thousands of templates, it is important to start with fragmented DNA of an appropriate size range. If the fragments are too large, they will not ligate efficiently into the vector, and the proportion of templates lacking inserts or containing short inserts will increase; even size-selected sonicated DNA contains a low proportion of very small fragments which ligate efficiently. This will increase the amount of sonicated DNA required and the total number of templates that need to be prepared. If the fragments are too small, the number of templates containing short or multiple inserts will increase in proportion. This too will increase the number of templates needed, and will also enhance the amount of effort required to resolve non-contiguous short sequences that ligate together and enter the database.

One of the major strengths of random sequencing is that each nucleotide is sequenced several times in order to obtain a single contiguous sequence containing all the overlapping gel readings. This redundancy is a key element in facilitating identification and removal of errors. Redundancies of six to ten are generally obtained. The number of templates required to complete a sequence depends on the redundancy required and the average length of sequence that can be read per clone. We usually sequence a number 30–40 times greater than the target size in kb.

The use of microtitre plates and electronic pipettes has greatly increased the rate at which templates can be prepared and sequenced, and form one aspect of automation [5]. Section 5.5 gives instructions for manually preparing the reactions for one sequencing gel using an electronic pipette with a single tip. Sections 5.6 and 5.7 include instructions for preparing reactions for a batch of several gels using an electronic pipette with eight tips. Increasing the size of the batch lengthens the sequencing process and sometimes reduces the quality of individual reactions, so each worker should find a suitable balance. We regularly prepare reactions for eight gels, and sometimes for larger batches.

The amount of information obtained per sequencing gel is maximized by using a buffer gradient system [10] in gels containing 96 lanes (24 templates). If samples are prepared beforehand, a single worker can comfortably process four gels each day. After electrophoresis, it is possible to dry gels directly without treating them in any way (see Section 6.7). The thickness of the gel, however, presents a barrier to emitted beta-particles and results in longer exposure times. We have been engaged in progressively reducing the amount of radioisotope in sequencing reactions, and, in accordance with earlier practice, now treat gels with acetic acid in order to fix the DNA and remove urea, thereby reducing their thickness when dried.

3.2   Compiling and Analysing Data

About half of the time involved in a sequencing project is used for reading autoradiographs, compiling the database, dealing with specific problems and analysing the completed sequence. Reading autoradiographs using a digitizer takes a major proportion of this time and great determination. It may take an hour of concentrated effort to read 24 templates. It seems likely that automated gel reading software will soon be acceptable for practical use; this will be a great advance.

We use Staden's sequence analysis package [11] operating in a MicroVAX II to assemble databases. Other options in software and hardware are available, but we have not tested them. We do not give full details of assembling databases in this chapter, but brief comments are made in Section 5.9. For analysing finished sequences, we use the software package of the Genetics Computer Group of the University of Wisconsin [12].

4.   MATERIALS

4.1   Equipment

Items are listed below in the order in which they are used in Section 5. Indication of specific suppliers does not imply that equipment from other sources is inferior.

Cup-horn sonicator (Heat Systems W-380; used at setting 7)

Agarose gel electrophoresis kit (Gibco-BRL)

Microcentrifuges (MSE Microcentaur)

15 mL polypropylene tubes (Falcon 2006)

Sterile 24-well flat-bottomed plates with lids (Nunc)

Pipetaid electric pipetting device (Scotlab)

Autoclaved cocktail sticks

Shaking benchtop incubator (Luminar IncShake used at setting 5)

Gilson racks for holding 80 microcentrifuge tubes (Scotlab)

Gilson Pipetmen P20, P200 and P1000 (Scotlab)

Rainin EDP-Plus multidispensing electronic pipette with 25, 100, 250 and 1000 µL liquid ends (Scotlab)

Water pump fitted to a Buchner flask, containing 50 mL of Stericol (Lever), attached to a 1 mL glass pipette with a blue tip on the end SMI mark III multitube vortexer (Alpha Laboratories; used at setting 6 for phenol extraction and at setting 4 for mixing solutions)

Sterile 96-well round-bottomed microtitre plates (Nunc)

Adhesive sealers for microtitre plates (Corning)

Rainin EDP-Plus M-8 multidispensing electronic pipette (eight tips, Scotlab)

Benchtop centrifuge (Beckman GPR)

Model S2 sequencing gel electrophoresis apparatuses with glass plates, 0.4 mm spacers and 0.4 mm double-fine sharkstooth combs (Gibco-BRL)

Sealing tape (Scotch brand 56)

3 µl sequencing gel loading pipette (Sigma; tip trimmed with a razor blade)

3000 V sequencing gel powerpacks (Pharmacia)

3MM chromatography paper (Whatman)

Gel dryers (Biorad model 583)

X-ray film (Kodak XS1) Computer-linked gel reading device (Summagraphics digitizer or a Science Accessories Corporation Grafbar Mark II)

5. METHODS

5.1   Generating Random DNA Fragments

5.1.1   Reagents

TE
  10 mM Tris-HCl (pH 8.0)
  1 mM EDTA
123 bp DNA ladder (Gibco-BRL)
50% (w/v) aqueous polyethylene glycol 6000 (PEG)
10 mg/mL ethidium bromide
1 M NaCl
95% (v/v) ethanol
10 x T4 DNA polymerase buffer
  330 mM Tris-acetate
  100 mM magnesium acetate
  660 mM potassium acetate
  5 mM dithiothreitol (DTT)
  pH adjusted to 7.9
  (supplied by Boehringer as restriction buffer A)
Stock nucleoside triphosphate solutions (Pharmacia): 100 mM dGTP,
dATP, dTTP, dCTP, ATP, 5 mM deaza-dGTP, ddGTP, ddATP, ddTTP,
ddCTP
2 mM dNTPs: 2 mM each of dGTP, dATP, dTTP, dCTP
T4 DNA polymerase (Boehringer; 3 units/µL)
Phenol/chloroform: melt phenol in a waterbath at 60°C and add
carefully to an equal volume of chloroform in a fumehood (take
appropriate safety precautions)
3 M sodium acetate (pH 5.5)
TE(0.1)
  10 mM Tris-HCl (pH 8.0)
  0.1 mM EDTA

5.1.2   Procedure

   1.   Place 1-5 µg of target DNA (see Section 3.1) in 200 µL of TE
        in a 1.5-mL microcentrifuge tube and sonicate for 30 s, using a
        cup-horn sonicator.

2. Electrophorese 10 μL of sonicated DNA on a 1.5% (w/v) agarose gel containing 0.5 μg/mL ethidium bromide, using a 123 bp ladder as markers. The DNA fragments detected by UV transillumination should range in size from about 250 to 3000 bp. If they are too large, sonicate again and check by gel electrophoresis.

3. Mix 172 μL of sonicated DNA with 64 μL of 50% PEG (i.e. to a final concentration of 8%) and 164 μL of 1 M NaCl. Incubate overnight in an ice bucket kept in a coldroom.

4. Pellet the DNA by microcentrifuging for 10 min. Transfer the supernatant to a fresh tube, add 500 μL of 95% ethanol to the pellet and microcentrifuge for 2 min. Discard the supernatant, wash the pellet again with 95% ethanol and dry it in a lyophilizer. Dissolve the DNA in 50 μL of water.

5. Electrophorese 5 μL of the PEG-precipitated DNA on a 1.5% agarose gel, using 10 μL of the remaining sonicated DNA and a 123 bp ladder as markers. The DNA fragments visualised should range in size from about 700 to 3000 bp. If the DNA is not precipitated by 8% PEG, add 8 μL of 50% PEG to the retained supernatant (thus increasing the concentration to 9%), mix and incubate on ice overnight. Then pellet and check the size of the DNA. If DNA of the appropriate size is still not precipitated, try increasing the concentration of PEG to 10%.

6. Mix 45 μL of DNA fragments with 5 μl of 10 x T4 DNA polymerase buffer, 5.6 μL of 2 mM dNTPs and 3 U of T4 DNA polymerase. Incubate at 37°C for 1 h. Extract with 100 μL of phenol/chloroform, add 6 μL of 3 M sodium acetate and 150 μL of ethanol and incubate on dry ice for 5 min. Pellet the DNA by microcentrifuging for 5 min and discard the supernatant. Add 500 μL of 95% ethanol, microcentrifuge for 2 min and discard the supernatant. Dry the pellet in a lyophilizer and redissolve it in 20 μL of TE(0.1).

5.2   Generating M13 Recombinants

5.2.1   Reagents

Bacteriophage M13mp19 RFI DNA (Gibco-BRL) which has been linearized with *Sma*I and treated with calf intestinal phosphatase

5 x DNA ligase buffer
  250 mM Tris-HCl (pH 7.6)
  50 mM MgCl^2
  5 mM DTT
  5 mM ATP
  25% (w/v) PEG 8000

T4 DNA ligase (Boehringer)

L-broth
  10 g/L bacto-tryptone (Difco)
  5 g/L yeast extract (Difco)
  5 g/L NaCl
  adjust pH to 7.0 and autoclave

L-broth agar
  15 g/L bacto-agar (Difco) in L-broth; autoclaved

  E. coli MAX efficiency DH5αF´IQ competent cells (Gibco-BRL)

Top agar
  6 g/L bacto-agar (Difco); autoclaved

IPTG: 30 mg/mL isopropyl β-D-thiogalactopyranoside (Gibco-BRL)

X-gal: 40 mg/mL 5-bromo-4-chloro-3-indolyl β-D-galactopyranoside
  (Sigma) in dimethylformamide (BDH)

E. coli XL1-Blue (Stratagene)

2YT broth
  16 g/L bacto-tryptone (Difco)
  10 g/L yeast extract (Difco)
  5 g/L NaCl
  adjust pH to 7.0 and autoclave

5.2.2 Procedure

1. Mix 2 µL (0.2 µg) of *Sma*I-cleaved, dephosphorylated M13-
   mp19 DNA, 2 µL of sonicated DNA fragments, 4 µL of
   5× DNA ligase buffer, 10 µL of water and 2 U of T4 DNA
   ligase. Also set up two control reactions omitting the DNA
   fragments or lacking ligase. Incubate overnight at room
   temperature. Add 80 µL of water to each reaction.

2. Melt 300 mL of L-broth agar in a microwave oven or boiling
   waterbath, allow it to cool to approximately 50°C on the bench
   and pour ten 90 mm L-broth agar plates. When they are set, dry
   them open and inverted in a 37°C room for 1 h.

3. Place 50 μL of competent *E. coli* cells in five Falcon 2006 tubes on ice. Gently add 2 μL of each of the three ligation reactions to separate tubes, and 2 μL of the M13mp19 RFI DNA supplied with the competent bacteria to the fourth tube. Do not add DNA to the fifth tube. Mix very gently and incubate the transfections on ice for 30 min.

4. Melt 100 mL of top agar and cool it in a 42°C water bath. Transfer 15 mL to five separate vessels in the water bath. Add 50 μL of IPTG, 100 μL of X-gal and 1 mL of a fresh overnight culture of *E. coli* XL1-Blue (grown in 2YT broth) to each. Incubate the transfections at 42°C for 1 min and then add them to the top agar.

5. Overlay the top agar mixtures onto L-broth agar plates (3 mL/plate) and allow the agar to set for 15 min at room temperature. Invert the plates and incubate them overnight in a 37°C incubator. We recommend five plates for the sonicated DNA ligated to vector and one plate each for the four controls.

6. The transfection efficiency will be indicated by the M13mp 19 RFI transfection. The ligated DNA transfection will give a mixture of blue and clear (recombinant) plaques, the DNA controls will give a modest number of blue plaques and perhaps one or two clear plaques, and the control lacking DNA will give no plaques.

7. Larger scale transfections (using 10 μL of DNA fragments ligated to vector and 200 μL of competent bacteria) may be carried out in order to obtain the required number of clear plaques. Plates can be stored inverted in a sealed box at 4°C for at least two weeks

5.3 Preparing Templates

5.3.1 Reagents

2 YT broth: see Section 5.2.1

PEG/NaCl
   20% (w/v) PEG
   2.5 M NaCl

TE, TE (0.1): see Section 5.1.1.

Phenol/TE
   Melt phenol in a waterbath at 60°C, equilibrate with TE at room

temperature, remove excess TE and store at -20°C (take appropriate precautions); when required, melt at 37°C and cool to room temperature
Sodium acetate/ethanol
Mix 10 mL of 3 M sodium acetate (pH 5.5) with 250 mL of ethanol

5.3.2   Procedure

1. Dilute 1 mL of a fresh overnight culture of *E. coli* XL1-Blue into each of two 100 mL bottles of 2YT broth. Aliquot 1.2 mL into each well of six 24-well plates using a Pipetaid fitted with a 10 mL pipette.
2. Visualize the bacterial plates on a lightbox and transfer one clear plaque to each well using cocktail sticks. Touching the bottom of the well with the stick will suffice for transfer. Fit lids to the plates and incubate in a shaking bench-top incubator at 37°C for 6 h. Humidify the incubator prior to and during incubation, using water-filled trays containing wicks.
3. Arrange 144 1.5 mL microcentrifuge tubes in three Gilson racks (see Section 6.2). Transfer 1 mL of culture to each tube, using a Pipetman P1000 and the same blue tip for all cultures. Pellet the bacteria by microcentrifuging for 5 min and tip the supernatants into clean tubes. Add 120 µL of PEG/NaCl to each tube using an EDP-Plus pipette, close the lids, mix by inverting several times and incubate overnight at 4°C.
4. Pellet the bacteriophage by microcentrifuging for 5 min and aspirate the supernatants using a water pump. Microcentrifuge the tubes briefly and remove the remaining supernatant, using a Pipetman P200 and the same yellow tip for all tubes.
5. Add 100 µL of TE and then 50 µL of phenol/TE to each tube using an EDP-Plus pipette (see Section 6.3). Place the racks of tubes in a multitube vortexer and mix vigorously for 1 min. Leave for at least 1 h and mix again for 1 min. Leave for at least 5 min and mix again for 1 min. Microcentrifuge the tubes for 2 min and transfer the supernatants to clean tubes using a Pipetman P200 and a fresh yellow tip for each. Add 250 µL of sodium acetate/ethanol to each tube using a Pipetaid fitted with a 5 mL pipette. Mix by inverting several times and incubate overnight at −20°C.
6. Pellet the DNA by microcentrifuging for 5 min. Aspirate the

supernatants using a water pump. Add 500 μL of 95% ethanol to each tube and aspirate, then aspirate the residual 95% ethanol. Dry the recombinant M13 templates by placing the racks of open tubes in a 37°C room for 1–2 h.

7. Add 30 μL of TE(0.1) to each tube using an EDP-Plus pipette. Dissolve the templates by shaking the racks of tubes at a moderate speed in a multitube vortexer for 20 s. Tap the racks on the bench a few times to bring down droplets. Transfer the templates to microtitre plates, using a fresh tip for each template. Cover the plates with adhesive sealers, number them and store at –20°C.

5.4 Annealing Primer to Templates

5.4.1 Reagents

TM
   100 mM Tris-HCl (pH 8.0)
   100 mM $MgCl_2$
Universal 17-mer primer: see Sections 2.3 and 6.4

5.4.2 Procedure

1. Thaw plates of templates and vortex them briefly. Place them in the standard orientation (see Section 6.2), and transfer 3 μL of each template to a fresh microtitre plate using an EDP-Plus M-8 pipette, so that the templates are arranged identically on the fresh plates. Wash out the tips after each set of eight templates by twice drawing up 100 μL of water from a 1 L beaker and expelling it into the beaker, absorbing the final small volume of expelled water on a tissue. Cover and store the remaining template stocks at –20°C.

2. For each pair of plates (192 templates), mix 2882 μL of water, 396 μL of TM and 22 μL of universal primer (see Section 6.4). Aliquot 205 μL of the mixture into each well of two rows of eight wells in a fresh microtitre plate. Transfer 15 μL of the mixture to each template well using an EDP-Plus M-8 pipette (see Section 6.3). Cover the plates, vortex briefly and heat in a 37°C incubator for 30 min (see Section 6.2). Annealed templates may be stored at –20°C.

5.5   Sequencing Batches of 24 Templates

5.5.1   Reagents

dNTP/ddNTP solutions: see Table 1 and Section 6.5

Table 1.   dNTP/ddNTP solutions[a].

	G	A	T	C
TE(0.1)	2000	1000	2000	2000
0.5 mM 7-deaza-dGTP	50	1000	1000	1000
0.5 mM dTTP	1000	1000	50	1000
0.5 mM dCTP	1000	1000	1000	50
5 mM ddGTP	20	—	—	—
5 mM ddATP	—	2	—	—
5 mM ddTTP	—	—	50	—
5 mM ddCTP	—	—	—	7

[a]Volumes are in µL

2'-deoxyadenosine 5'-[α-³⁵S]thiotriphosphate ([³⁵S]dATP) (SJ264; Amersham)

0.1 M DTT (dithiothreitol)

Klenow fragment of *E. coli* DNA polymerase I (see Sections 2.3 and 6.4)

Chase solution
   0.125 mM each of dGTP, dATP, dTTP and dCTP

5.5.2 Procedure

1. A batch of 24 templates can be electrophoresed on a single gel. Using a marker pen, divide a fresh microtitre plate into three sections (see Section 6.2). Each will accommodate the reactions for one row of templates. Dispense annealed templates from three rows of eight wells by taking up 8 µL into an EDP-Plus pipette and dispensing 2 µL (see Section 6.3) into four wells in a column.

2. Place 50 µL of the dNTP/ddNTP solutions (see Table 1 and Section 6.5) into four tubes. Add 2 µL of 0.1 M DTT, 2.5 µL

(25 µCi) of [35S]dATP and 0.5 µL of Klenow fragment (or a larger volume of Klenow fragment diluted in TE if necessary; see Section 6.4) to each tube. Mix and centrifuge briefly. Using an EDP-Plus pipette, aliquot 2 µL of G mixture onto the sides of the wells in the first row of each section. Repeat this process for the A, T and C solutions and the other three rows of each section, using a fresh tip for each mixture. Cover the plate, centrifuge briefly in a benchtop centrifuge and vortex briefly. Transfer immediately to a 37°C incubator for 10 min.

3. Add 2 µL of chase solution to the side of each well using an EDP-Plus pipette. Cover the plate, centrifuge briefly in a benchtop centrifuge and vortex. Transfer the plate immediately to a 37°C incubator for 10 min. Store at –20°C.

5.6 Sequencing Batches of 192 Templates

5.6.1 Reagents

As for Section 5.5.

5.6.2 Procedure

1. A batch of 192 templates gives material sufficient for eight sequencing gels. Mark eight fresh microtitre plates as described in step 1 of Section 5.5 and number them. Dispense the annealed templates in batches of eight by taking up 12 µL into an EDP-Plus M-8 pipette and dispensing 2.5 µL (see Section 6.3) into the four rows of a section. Wash the tips twice with 100 µL of water between sets of templates.

2. Place 500 µL of the dNTP/ddNTP solutions (see Table 1 and Section 6.5) into four tubes. Add 20 µL of 0.1 M DTT, 25 µL(250 µCi) of [³⁵S]dATP and 5 µL of Klenow fragment to each tube. Mix and centrifuge briefly. Complete step 2 of Section 5.5, but dispense 2.5 µL, rather than 2 µL, aliquots into the wells.

3. Dispense 300 µL of chase solution into each of eight wells of a microtitre plate and add 2.5 µL directly into the reactions using an EDP-Plus M-8 pipette. Press the trigger before touching the reaction mixes with the expelled drops of chase solution to avoid transferring reaction products from well to well. Cover

the plates, vortex briefly and transfer them immediately to a 37°C incubator for 10 min. Freeze the plates on a tray of dry ice and store them at −20°C.

5.7   Sequencing Larger Batches of Templates

1.   Larger batches of 394 (and even up to 576) templates, which give material sufficient for 16 (or 24) sequencing gels, may be prepared. Since the time taken to process large numbers of reactions is excessive in comparison with incubation times, a large batch may be separated into four parts by altering the arrangement of templates in the microtitre plates. Using step 1 of Section 5.6 as a guide, aliquot from each plate of annealed templates into four fresh microtitre plates to replicate the original plate (i.e. dispense the first row of templates into the first row of each of the four plates, and so on). Mark the plates G, A, T and C and number them appropriately.

2.   Using step 2 of Section 5.6 as a guide, prepare the sequencing mixtures. 1 mCi of [^{35}S]dATP will suffice for 394 templates, but it may be necessary to decrease the amount of ddATP in the A solution (see Table 1) and to expose the autoradiographs for longer. Divide each sequencing mixture equally between eight wells in a fresh microtitre plate. Using an EDP-Plus M-8 pipette, aliquot 2.5 µL of G mixture directly into the wells in the G plates. Cover the plates, vortex briefly and transfer them immediately to a 37°C incubator for 10 min. Using step 3 of Section 5.6 as a guide, add chase solution, incubate and freeze. Repeat this process separately for the A, T and C plates.

5.8   Gel Electrophoresis

5.8.1   Reagents

Sigmacote (Sigma)

10 × TBE
   109 g/L Tris
   55 g/L boric acid
   9.3 g/L EDTA

Acrylamide solution

40% (w/v) acrylogel 5 premix (BDH); deionized

TGM

Add 6 mL of 10 × TBE and 18 mL of acrylamide solution to 55.2 g of urea

Increase the volume to about 110 mL with water

Dissolve by stirring with careful heating

Adjust the final volume to 120 mL with water and filter the solution into a flask on ice

This gives sufficient solution for two sequencing gels

BGM

460 g/L urea

50 g/L sucrose

50 mg/L bromophenol blue

250 mL/L 10 × TBE

150 mL/L acrylamide solution

filter and store at 4°C in the dark

APS: 25% (w/v) ammonium persulphate

TEMED: N,N,N′,N′-tetramethylethylenediamine

Formamide-dyes

1 g/L xylene cyanol FF

1 g/L bromophenol blue

10 mM EDTA in deionized formamide

10% (v/v) acetic acid

5.8.2 Procedure

1. It is convenient to prepare gels in pairs. Clean siliconised sequencing gel plates by polishing with a small volume of 95% ethanol, and assemble the gel sandwiches using sealing tape (see Section 6.6). The sandwiches should have two large foldback clips on each side, positioned over the spacers. Position them vertically.

2. Measure 60 mL of freshly prepared TGM into one beaker and 12 mL of BGM into another. Add 24 μL and 120 μL of APS to the BGM and TGM, respectively. Add the same volumes of TEMED and mix. Use a Pipetaid to take up 12 mL of polymerising TGM into a 25 mL glass pipette and then slowly take up all the polymerising BGM, allowing a few air bubbles

to disturb the interface. Using a single side-to-side movement, dispense about 12 mL of the gradient into the sandwich and repeat the movement with the remaining 12 mL. Take up as much of the remaining TGM as possible into the pipette and dispense using the same movement. Strike the glass plates vigorously to dislodge air bubbles, then lay the sandwich almost horizontally on a low object such as a tube rack. Insert two adjacent combs into the top 5 mm of the sandwich, teeth uppermost, and clamp them with three large foldback clips.

3. Pour a second gel as described in step 2.

4. After allowing the gels to set for 15–30 min, remove the clips and the tape around the bottom and place the sandwiches in sequencing apparatuses. Add 500 mL of 0.5 × TBE to each top and bottom buffer compartments (see Section 6.6). Use a syringe needle to remove excess polyacrylamide around the combs, and the points of a pair of scissors to remove the combs gently, taking care not to damage the gel surface. Wash the tops of the gels using a 60 mL syringe and needle. Wash the gel surfaces a second time while the samples are boiling (see step 5) and again immediately prior to replacing the combs and loading.

5. Using an EDP-Plus pipette, add 2.5 μL of formamide-dyes to the side of each well of two plates of sequencing reactions immediately after removal from the –20°C freezer. Cover the plates, tap the formamide-dyes to the bottom of the wells and vortex briefly. Place the trays one at a time in a boiling water bath for 1 min and cool them on ice for 5 min. Replace the combs in the surface of the gel with teeth downwards. Load 2 μL of sample in the order GATC for each template. Electrophorese the two gels at a combined power of 100 W using a single power pack (see Section 6.6). If the samples were prepared as in Section 5.7, it is necessary to set up and load four gels at once, since the four sets of reactions are on different plates.

6. After 2–2.5 h the bromophenol blue dye in the samples will be at the bottom of the gels. Turn off the power, remove the sandwiches from the apparatuses and discard the tape. Gently ease off one glass plate from each gel using sharp scissors applied into the bottom of the sandwich and remove the combs

and spacers. Place the plate containing the gel on a tray in a fumehood and cover the surface with as much 10% acetic acid as possible without allowing it to flood off the plate. After 5–10 min, tip the surface liquid into the tray, rinse with 10% acetic acid and apply more. Repeat this step once again. Place two sheets of 3MM paper on the gel, invert the assembly and remove the glass sheet. Cover the gel with clingfilm, trim excess paper and clingfilm and dry the gel on a gel dryer for 45 min at 80°C.

7. Remove the clingfilm and the second sheet of paper from the dried gel. Expose the gel to X-ray film in a cassette overnight (to check whether the sequencing reactions have worked) or for about a week (to obtain archive autoradiographs).

5.9   Compiling the Database

1. Label the autoradiographs according to template number on the microtitre plates, and read them using a computer-linked gel reading device. We are usually able to read about 270 nucleotides per clone, and on rare occasions over 350 nucleotides can be read.

2. Assemble the database using Staden's sequence analysis package (SAP). Take appropriate steps if the target contains separate repeat regions, since data from these will condense into a single overlapping sequence. If the target contains highly reiterated regions, it may be necessary to carry out specific experiments to sequence through them. Contaminating sequences usually arise from cellular DNA, and overlap neither the target nor each other. They should be removed from the database once the target is contiguous.

3. Print the database and edit it by making careful reference to the autoradiographs, endeavouring to read every base from the best autoradiographs on both strands. Highlight regions of difficulty. Transfer changes to the database using SAP. Only the consensus needs to be accurate - it is not necessary to make every nucleotide agree with the consensus.

4. Resolve regions of difficulty by performing additional sequencing reactions with selected templates (plaque-purified by retransfection if necessary), using the universal primer or cus-

tom primers. In the last resort, use a thermostatically heated gel apparatus to resolve compressions.

6. PRACTICAL POINTS

6.1 Hazardous Reagents

The methods involve the use of several hazardous or potentially hazardous reagents, including recombinant bacteria, ethidium bromide, phenol, acrylamide, N,N'-methylenebisacrylamide, UV radiation and radioactive isotope. It is essential to follow precautions for handling and disposing of such substances as advised by the relevant national or local authorities.

6.2 Handling Microcentrifuge Tubes and Microtitre Plates

Microcentrifuge tubes should be arranged in three rows of 16 per Gilson rack. Since the templates are random with respect to the target sequence, it is not necessary to label them or the racks at any stage or to place them in microcentrifuges in any predetermined order. It aids aspiration of supernatants, however, to use a standard tube orientation in the microcentrifuge, so that the bacteriophage or template pellet is on the same side of every tube. All microcentrifugation steps are carried out at maximum speed (13 000 r.p.m.) at room temperature. Tubes should be placed in the microcentrifuge in groups of two or three rather than singly. As many of the steps as possible, including aspirating supernatants, should be carried out without lifting tubes from the racks, and the tube lids should be opened or closed only when necessary.

The standard orientation of a microtitre plate is with eight *columns* along the top and twelve *rows* down the side. A subdivision of four *rows* of eight columns is a *section*. Annealed templates and sequencing reactions may be set up in large batches and stored at –20°C; the former for months and the latter for days. When incubating reactions in a 37°C incubator, it is advisable to place plates on a solid shelf to allow rapid temperature equilibration.

6.3 Use of Multidispensing Electronic Pipettes

Both types of EDP pipette used are capable of multidispensing (i.e. taking up a larger volume and dispensing equal aliquots with smaller volumes). This mode should be used where appropriate. The EDP-Plus pipette dispenses 2 μL volumes accurately (see Section 5.5), but the smallest volume that can be dispensed reproducibly using the EDP-Plus M-8 pipette is 2.5 μL (see Section 5.6). The tips on the EDP-Plus M-8 pipette are not changed at all during the procedures.

6.4 Home-made Reagents

Universal primer and Klenow enzyme are prepared in large concentrated batches (see Section 2.3). The volumes that we are currently using are given in step 2 of Sections 5.4, 5.5 and 5.6. It is necessary, however, to optimize the amounts of these reagents. Our stock solution of primer is approximately 5 mM, and the working solution (as used in step 2 of Section 5.4) is 50 μM. Our stock solution of Klenow enzyme has an activity of approximately 20–50 units/μL.

6.5 dNTP/ddNTP Solutions

The precise compositions depend on the G+C content of the target DNA, the actual concentrations of stock dNTP and ddNTP solutions and the concentration of [^{35}S]dATP used. After obtaining them empirically on a small scale (see Section 5.5), make up sufficient solutions for the project and store them at –20°C.

6.6 Gel Electrophoresis

Glass plates are prepared by washing with soapy water, rinsing with water, drying, polishing with a small volume of Sigmacote in a fume hood, washing with 95% ethanol and then polishing with a small volume of 95% ethanol. After several uses, gels start to adhere to both glass plates after electrophoresis, and the plates need to be resiliconised. Gels can be stored overnight before use the next day if they are placed in the sequencing kits without removing the bottom tape and combs, and buffer is added only to the top compartment. Gels may be electrophoresed singly at 60–70 W.

6.7 Drying Gels

Gels may be dried without fixing in acetic acid. Since electrophoresed DNA seems to concentrate against the longer (cooler) plate, the gel must be obtained on this plate and then transferred to paper. On some occasions the gel comes off on the shorter plate. If this occurs, the gel must be inverted. To do this, transfer the gel to a sheet of paper, place two sheets of paper on the exposed surface of the gel and invert the assembly. Soak the upper single sheet of paper with 10% acetic acid and remove it. Cover the gel with clingfilm and dry it. Gels may be stored on paper for several hours before drying without loss of resolution. Unfixed gels should be dried for 1.5 h at 80°C.

6.8 Troubleshooting

The commonest cause of apparent variability in template quality is inadequate phenol extraction of pelleted bacteriophage. This is manifested on autoradiographs as variation in the intensity of products from different templates in a batch. It is important to follow step 5 of Section 5.3 carefully; further vortexing can be included if necessary. Reference [13] gives an excellent account of how to tackle the majority of problems that are encountered in sequencing.

ACKNOWLEDGMENTS

We thank Duncan McGeoch and Barbara Barnett for criticism of the manuscript. A.J.D. is a member of staff of the MRC Virology Unit.

E.A.R.T. is supported by the Equine Virology Research Foundation

REFERENCES

[1] Sanger, F., Nicklen, S. and Coulson, A.R. DNA sequencing with chain-terminating inhibitors. Proc. Natl. Acad. Sci. USA, 74 (1977) 5463–5467.

[2] Messing, J. A multi-purpose cloning system based on the single-stranded DNA bacteriophage M13. In *Recombinant DNA Technical Bulletin*, NIH Publication No. 79-99, National Institutes of Health, Bethesda, 1979, No. 2, pp. 43–48.

[3] Oliver, S.G., van der Aart, Q.J.M., Agostoni-Carbone, M.L., Aigle, M.L.,

Alberghina, L. et al.. The complete DNA sequence of yeast chromosome III. Nature, 357 (1992) 38–46.

[4] Messing, J. and Bankier, A.T. The use of single-stranded DNA phage in DNA sequencing. In Howe, C.J. and Ward, E.S. (eds.) *Nucleic Acids Sequencing: A Practical Approach*. IRL Press, Oxford, 1989, pp. 1–36.

[5] Bankier, A.T. and Barrell, B.G. Sequencing single-stranded DNA using the chain-terminating method. In Howe, C.J. and Ward, E.S. (eds.) *Nucleic Acids Sequencing: A Practical Approach*. IRL Press, Oxford, 1989, pp. 37–78.

[6] Joyce, C.M. and Grindley, N.D.F. Construction of a plasmid that overproduces the large proteolytic fragment (Klenow fragment) of DNA polymerase I of *Escherichia coli*. Proc. Natl. Acad. Sci. USA, 80 (1983) 1830–1834.

[7] Davison, A.J. Experience in shotgun sequencing a 134 kilobase pair DNA molecule. DNA Sequence, 1 (1991) 389–394.

[8] Davison, A.J. Channel catfish virus: a new type of herpesvirus. Virology, 186 (1992) 9–14.

[9] Telford, E.A.R., Watson, M.S., McBride, K. and Davison, A.J. DNA sequence of equine herpesvirus-1. Virology, 189 (1992) 304–316.

[10] Biggin, M.D., Gibson, T.J. and Hong, G.F. Buffer gradient gels and 35S label as an aid to rapid DNA sequence determination. Proc. Natl. Acad. Sci. USA, 80 (1983) 3963–3965.

[11] Staden, R. Computer handling of DNA sequence projects. In Bishop, M.J. and Rawlings, C.J. (eds.) *Nucleic Acid and Protein Sequence Analysis: A Practical Approach*. IRL Press, Oxford, 1987, pp. 173–217.

[12] Devereux, J., Haeberli, P. and Smithies, O. A comprehensive set of sequence analysis programs for the VAX. Nucleic Acids Res., 12 (1984) 387–395.

[13] Ward, E.S. and Howe, C.J. Troubleshooting in chain-terminating DNA sequencing. In Howe, C.J. and Ward, E.S. (eds.) *Nucleic Acids Sequencing: A Practical Approach*. IRL Press, Oxford, 1989, pp. 79–97.

Chapter 11

PULSED-FIELD GEL ELECTROPHORESIS
APPLICATIONS TO BACTERIA

Peter R. Stewart, Wafa El-Adhami and
Barbara Inglis

OUTLINE

Methods in Gene Technology, Volume 2, pages 177–205
Copyright © 1994 JAI Press Ltd
All rights of reproduction in any form reserved.
ISBN: 1-55938-264-3

1. INTRODUCTION

The capacity to resolve large fragments of DNA on electrophoretic gels using pulsed electric fields, first described by Schwartz and co-workers in 1983 [1], has been an advance in DNA technology of remarkable significance. It has offered the means of resolving and isolating fragments of DNA 100 times or more greater than the size manageable using continuous or steady field electrophoresis (which is typically 50 kb or less). This has meant that for bacterial genomes, chromosome sizes could be measured directly, genes on fragments of 100–1000 kb could be examined for physical linkage, and entire genomes could be compared directly by endonuclease mapping and by duplex hybridization of individual chromosomal fragments with one another. As Smith and colleagues [2] point out, "there is a size range between 100 kb and 2000 kb which is too large to approach by standard molecular techniques and too small to resolve in cytogenetic and linkage analysis". PFGE neatly fills this gap.

This size range is also of great interest to the bacterial geneticist because bacterial genomes range in size from about 500 kb to 10 000 kb [3], and fragments of such genomes in the range 50–1000 kb are readily generated using rare-cutting restriction endonucleases [see Chapter 12].

PFGE is thus a technique of central importance in the mapping and sequencing of entire genomes, prokaryotic or eukaryotic.

2. TYPES OF PULSED-FIELD ELECTROPHORESIS

The terms *continuous* or *steady-field gel electrophoresis* refer to conventional methods of DNA fragment or molecule separation where the electric field direction and strength remain constant during the separation.

The term *pulsed-field gel electrophoresis*, or PFGE, is a generic term referring to gel electrophoresis systems in which the electric field is varied, usually in a controlled, pulsatile fashion. The variation is applied most commonly to field direction, usually some change of angle between 90° and 180° around the field axis, but the pulsing may also be in field strength (volts per cm through the gel matrix). In the case of a field direction change of 180°, a reversal of polarity of the field is implied; this type of electrophoresis, the simplest type of PFGE, is referred to as *field-inversion gel electrophoresis*, or FIGE [4] (Figure 1*a*).

Orthogonal field-alternation gel electrophoresis, or OFAGE, is PFGE in which the direction of the field varies, at some predetermined angle between 90° and 140° approximately (Figure 1*b*). OFAGE was the first type of PFGE described [5,6], and because of the geometry of the point or strip electrodes used, the electric field is not homogeneous. This remains a characteristic of OFAGE systems, and as a consequence the molecules do not necessarily migrate with a linear trajectory through the gel matrix. Lane widening may also occur leading to band sharpening and thus better resolution of fragments of similar size. Against this, however, is the difficulty of directly comparing the extent of migration (and thus the size) of DNA molecules that are not in immediately adjacent lanes.

Two variants of OFAGE are *contour-clamped homogeneous-field gel electrophoresis*, or CHEF [7] and *pulsed homogeneous-orthogonal gel electrophoresis*, or PHOGE [8]. In these, the electrode configuration and the electric potential switching are controlled in such a way as to generate electric fields which are homogeneous. In CHEF, the electrode arrays are hexagonal (Figure 1*c*) giving a field orientation which is pulse-switched by 120°, and in PHOGE the array is square giving 90° field switching. Both have the advantage over OFAGE that the DNA molecules migrate with linear trajectories, similar to those seen in continuous field electrophoresis and FIGE. This means that fragment sizes can be compared accurately across multiple lanes in a gel so that size identity can be established, or specific DNA molecules isolated by cutting gel bands from individual lanes. It also allows the accurate indexing of gel bands to transferred fragments on membrane filters in DNA hybridization analysis. CHEF is the most common type of PFGE now used.

Programmable autonomously-controlled electrode gel electro-phoresis, or PACE, is a variant of CHEF utilising an electrode geometry which allows the field angle to be varied over a wide range [9].

Other variants of OFAGE include *rotating-field gel electrophoresis,* or ROFE [10], in which the electrode array is mechanically rotated around the gel (Figure 1*d*), and *transverse alternating-field*

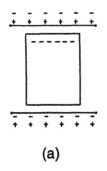

(a)

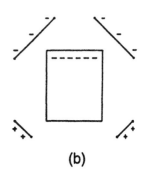

(b)

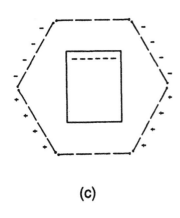

(c)

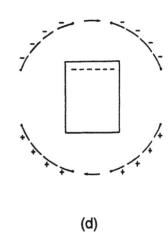

(d)

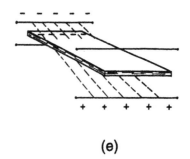

(e)

electrophoresis, or TAFE [11], where the electrodes are set out of the plane of the gel (Figure 1*e*).

Zero-integrated field electrophoresis, or ZIFE [12], is a variant of FIGE in which backward and forward pulse periods are different (as in FIGE), but the product of pulse time and field strength ($T_p \times E$) in both directions is constant. Thus, for example, if the reverse pulse time is one third that of the forward pulse, the field strength in the reverse direction is three times that in the forward. In this case DNA molecules 'drift' down the gel because of asymmetric conformational and matrix binding effects on the DNA molecules at high and low field strengths, as described below.

Figure 1. Types of pulsed-field gel electrophoresis
(*a*) Field-inversion gel electrophoresis (FIGE) involves a configuration of gel slab and electrodes the same as that for constant field (conventional) agarose gel electrophoresis. The electrode polarity is reversed at intervals to generate a 180° change (inversion) in field direction. (*b*) Orthogonal field-alternation gel electrophoresis (OFAGE) requires sets of asymmetric electrodes placed at approx. 60° to the axis of the gel slab. Alternating the field between the sets of electrodes results in an overall field direction change of approx. 120°. However, the field intensity and direction is not uniform across the gel slab, resulting in non-linear trajectories (lanes) for the DNA fragments in their migration through the gel. These trajectories are more distorted (but are symmetrical) as distance increases away from the mid-line of the gel. (*c*) Contour-clamped homogeneous-field gel electrophoresis (CHEF) involves a series of electrodes arrayed hexagonally around the gel slab and a complex switching device coordinating field intensity changes between anodic and cathodic pairs. The outcome is a 120° change in field direction as in OFAGE, but the programming of the field intensity changes with field direction results in a uniform field intensity within the electrode perimeter. Pulsed-homogeneous orthogonal-field gel electrophoresis (PHOGE) is a variant of CHEF in which the electrode perimeter is square or rectangular, and field direction alternation is 90°. (*d*) Rotating-field gel electrophoresis (ROFE) involves the mechanical rotation of the same electrode pair in relation to the axis of the gel slab. (*e*) Transverse-alternating field electrophoresis (TAFE) has two sets of electrodes set at alternate positions outside the plane of the gel slab.

3. THEORETICAL CONSIDERATIONS

In considering the behaviour of DNA molecules in electric fields, it should be noted that a 1000 kb (1 Mb) DNA molecule approximates a very thin, very long, highly flexible filament, with the approximate physical dimensions of 2 nm thickness by 300 000 nm length, or an axial ratio of 150 000. Such molecules are exceedingly fragile in their susceptibility to shear forces. Yet when bacterial cells are lysed *in situ* in the gel matrix (Section 5), it may be necessary to deliberately linearise the circular bacterial chromosome and plasmids in order for the molecule to be released from the lysing gel plug to migrate into the electrophoresis gel proper. Circular molecules have quite different migratory properties in gel matrices, and supercoiled circular molecules behave differently from relaxed circles. Linearization is thus essential when attempting to compare DNA molecule sizes. Gamma radiation can be used in a controlled way to achieve linearisation by strand breakage [13].

The movement of long, thin, flexuous molecules through a gel matrix with pore dimensions of 80–120 nm [14] in pulsatile electric

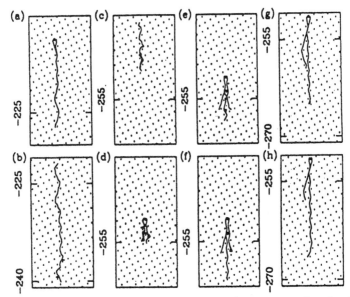

Figure 2. Variation in the conformation of a DNA molecule during agarose gel electrophoresis. (From Deutsch and Madden [16], with permission.)

fields is a process for which a theoretical framework has long been sought. However, a consistent and generally acceptable theoretical basis for DNA behaviour during PFGE has yet to be described. Two conceptual frameworks have been laid out. In the first, the DNA molecules are proposed to adopt a predominantly elongate state, and to reptate through 'tubes' in the gel matrix. In the second, DNA molecules have substantial random structure, including folding, kinking, coiling and knotting; these molecules migrate through a gel matrix of random point obstacles, fibres or pores. In the first case the DNA molecules are retarded by frictional effects in the tubes, and in the second by hooking around and pinning against the gel obstacles. The pulsed alternation of field direction permits the DNA molecules to reorientate, and new pathways are presented which permit migration to continue. Since reorientation time is a function of the size of the molecule, longer pulse times are required to resolve larger molecules.

The two models accurately predict different aspects of the behaviour of large DNA molecules in pulsed-field electrophoresis, but neither accounts for all aspects of the process.

The model of DNA behaviour in pulse fields outlined by Deutsch and colleagues [15–17] provides an intuitively satisfying explanation, though it is unlikely that this will be the last word on the matter. It is given here not as any judgment of the validity of other models but because it provides for the non-expert some idea of the complex behaviour of DNA molecules in an agarose gel matrix. In this model, Deutsch and his co-workers propose that DNA molecules do not remain in an extended state, but due to entanglement with gel fibres, or variable migration velocity along the length of the DNA molecule (the leading end moving more slowly than the middle or trailing end, for example), kinking, looping, and bunching of the molecule occurs (Figure 2). The development of this non-linear structure determines the rate of progress of the molecule through the gel matrix and, since the extent of kinking, looping and bunching is a function of chain length (thus molecular size), differential rates of migration through the gel result. This would also be the case in a steady electric field, but with the field acting continuously in one direction, large molecules would ultimately become entrapped against and around gel particles or fibres, and thus not migrate far through the gel or smear out along it. The effect of field pulsation, particularly with a changing directional component, is to permit new configurations of the entrapped DNA molecules to form, permitting release to occur, and new migratory routes to be tested. The molecules thus keep moving, and since the net movement is still a function of non-linear structure, the migration rate remains a function of molecular size, and molecules of identical size migrate at the same rate because entrapment of molecules, as would occur under continuous field conditions, is minimised.

As noted earlier, pulse durations in the two field directions must be sufficiently long that they permit the molecules to assume new non-linear configurations characteristic of that particular set of field and matrix conditions. Shorter pulse times result in the molecules undergoing little net movement. Despite what seems an obvious relationship between molecule size and reorientation time, this cannot be a complete explanation of the effects of pulse time, migration and resolution of DNA molecules. When high frequency pulses are superimposed on the longer pulses in a ZIFE system (the high frequency pulses are typically 5–10% of the normal pulse time, but with reversed polarity), molecular trapping is apparently decreased; narrower bands consequently result and electric fields can be increased leading to shorter electrophoresis times [18].

Deutsch and his colleagues also propose that DNA knotting and knot tightening (again, functions of molecular size) are important in determining the migration of large DNA molecules. Tight knotting would not be readily reversed, and molecules so affected would not be greatly responsive to changes in field direction. This would serve to deplete the larger molecules preferentially from a mixture, and may account for the DNA which commonly does not migrate out of the lysing blocks at the electrophoretic origin.

4.   IMPORTANT FACTORS IN THE PRACTICE OF PFGE

Regardless of the uncertainties which beset the theoretical underpinnings of PFGE, for the pragmatic, practising biologist the important fact is that PFGE in its various manifestations works, and can be made to work very well.

In setting up a PFGE system to examine large DNA molecules or fragments, there are five major practical determinants to be considered in optimising the resolution of large DNA molecules on gels. They are:

- field strength
- pulse time
- alternating field angle
- gel concentration (and agarose quality)
- temperature

These parameters are interdependent and require careful exploration to generate a successful PFGE system. Fortunately, a great deal of this exploratory analysis has already been done and is reported in accounts of PFGE applied to a wide and growing range of bacteria (Section 7). However, it is worth briefly stating the functional significance of variation of these parameters, as a guide to what might be expected to happen in a practical sense should they vary away from the optimum.

Field strength (measured as $V cm^{-1}$ or voltage drop per unit distance between electrodes) determines the rate of migration of charged molecules through a gel matrix, and hence maximum field strength would appear to be desirable. However, higher voltages mean larger current flow in the buffer and gel, which leads to heating. If the heat sink parameters (buffer circulation rate, refrigeration capacity, heat transfer capacity) are not adequate, increases in temperature and local heating effects may lead to band or lane distortion. Field strength is also linked to molecular relaxation rate, that is the rate at which DNA molecules adopt new configurations in response to altered field direction. In general, as field strength increases the pulse time needed for resolution of different size molecules should decrease. However, knotting and irreversible pinning of molecules to gel matrix fibres occurs at high field strength; resolution, particularly of larger molecules, is then impaired.

Pulse time should be greater for larger molecules, for reasons set out earlier; as a starting point from which to explore individual experimental protocols, pulse times of 1–2 s for <50 kb, 10 s for 50–200 kb, 60 s for 200–800 kb, 120 s up to 1400 kb have been suggested by Smith *et al.* [2].

Increased pulse time, like increased field strength, may result in increased molecule trapping, and thus smearing and poor resolution of bands on gels. But long pulses are needed to resolve larger molecules. Optimization of field intensity (E) and pulse time (T_p) have been

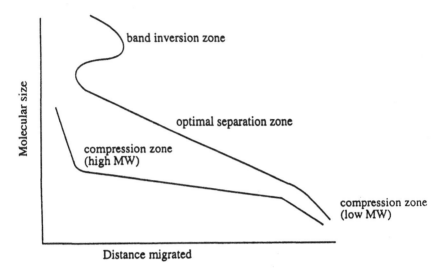

Figure 3. Idealised relationship between molecular size and gel migration in FIGE (upper) and CHEF (lower) systems. (After Carle *et al.* [4], Bostock [88] and unpublished observations.)

described as related by a constant, W (the 'window function'), where $W = E^{1.4} \times T_p$ for molecules larger than 1 Mb [13]. Once W is characterised for a particular electrophoretic system, it provides a convenient measure of how speed of separation (higher values of E) can be traded off against T_p.

Reorientation angles (alternating field direction) should be somewhere between 90° and 180°, with 120° a common setting. In many systems this is an invariant parameter.

The *properties of the gel matrix* appear to have received little attention, either in theoretical or practical terms. The heterogeneity of particle or pore size in agarose gels, charge density (which is negative and may thus determine DNA repulsion by the matrix and thus affect frictional effects), and chemical purity are probably important. Gel concentrations used are similar to those used in steady-field gel electrophoresis, approx. 0.8–1.2% w/v. Band sharpness is often greater at the higher concentrations, but there seems to be little further improvement beyond about 1.2%. Quality (and consistency of quality) of agarose is important. The best grades available should be used, and testing of different brands is worthwhile to identify those which give the best separations, sharpest bands, and most consistent results between batches.

Temperature influences the rate at which DNA molecules relax and adopt new configurations when the electric field direction changes. Temperatures of 12–14°C appear to be optimal in increasing relaxation rates while at the same time avoiding band and lane distortion.

Idealized relationships of fragment size and mobility for FIGE and CHEF systems are shown in Figure 3. The important features are that in FIGE systems ambiguity is likely to occur with large molecules since migration does not show an approximately inverse relationship with molecular size, and in both systems there may be regions of compression of mobility as a function of size, seen as a steepening of the curve relating the two. Other PFGE systems also exhibit this latter characteristic, and it is important that the operating parameters for any new geometrical configuration or regime of electric field pulsation be explored in order to determine the nature of such divergences from those anticipated between mobility and molecular size.

5. METHODS

5.1 DNA Preparation and Electrophoresis

The following sections describe the methods followed in our laboratory (modified from the method of [19] for *Escherichia coli*) for the preparation of intact genomic DNA of *Staphylococcus aureus* and other staphylococci for electrophoresis in FIGE [20] or CHEF [21, 22] systems.

5.1.1 Buffers and Reagents

ST buffer
   1 M NaCl
   10 mM Tris-Cl pH 7.6
Low melting temperature agarose (e.g. Sea Plaque from FMC)
   For making plugs, use 1% solution in TE buffer
TE buffer
   10 mM Tris-Cl pH 7.6
   1 mM sodium EDTA pH 7.5

Lysis solution
   1 M NaCl
   100 mM EDTA pH 7.5
   6 mM Tris-Cl pH 7.6
   0.5% Sarkosyl
   0.2% deoxycholate
   20 mg mL^{-1} DNAase-free RNAase
   1 mg mL^{-1} lysozyme
   50 µg mL^{-1} lysostaphin
The enzymes are added from concentrated stocks just prior to use
ES
   0.5 M sodium EDTA pH 9.3
   1% Sarkosyl
ESP
   ES + 0.5 mg mL^{-1} proteinase K
Electrophoresis gels
   normally 1% type II agarose, medium EEO, Sigma or equivalent,
   in 0.5 × TBE
TBE
   90 mM Tris-base
   90 mM boric acid
   2.5 mM EDTA acid, pH 8.2
   Use at half strength for preparation of running gels

5.1.2   Preparation of Intact Genomic DNA from Staphylococci

For effective application of PFGE to genomic DNA from bacteria, it is essential that the DNA be isolated from cells under conditions which prevent damage to the DNA by physical shearing or non-specific enzymic cleavage. Cells are therefore lysed and other cellular macromolecules (proteins, RNA, lipids) are degraded or solubilized by enzymic digestion and detergent treatment directly in agarose gel plugs which can then be inserted intact into the loading wells of the pulsed-field gel [5].

All reagents used should be of analytical or molecular biological grade to avoid problems caused by DNAases or inhibitors. In addition, sterile equipment is used and gloves are worn throughout the preparation process. An overnight shaken broth culture of *S. aureus* (grown on trypticase soy broth (TSB), Oxoid) is inoculated into fresh TSB (1 mL into 100 mL) and incubated shaken at 37°C until the A_{600}

reaches 0.15 (approx. 1.5×10^8 cells mL^{-1}). This takes approximately 1 h but will vary between strains and species. The cells synchronously enter exponential growth phase and undergo one to two rounds of chromosome replication. Exponentially growing cells are more readily lysed to release their DNA. However, workers using other bacteria have described the use of stationary phase broth-grown cells [23,24]. Some workers apply a growth inhibitor block (e.g. a protein synthesis inhibitor such as chloramphenicol [25]) to permit completion of synthesis of partially replicated chromosomes. It is worth testing for the most suitable growth conditions when beginning work with a new organism.

Broth cultures or cell suspensions are chilled on ice, centrifuged at 10 000 g for 15 min at 4°C, resuspended in 10 mL of ST and repelleted. The cell pellet is resuspended in 1.9 mL of ST to give 8×10^9 cells mL^{-1} (if A_{600} was not 0.15 in the subculturing step described earlier, then the volume (mL) of ST to resuspend the cells is calculated as $A_{600} \times 13$). This will give 3×10^8 cells, or approximately 1 µg DNA, per agarose half-plug (approx. 75 µL, see below). Estimating the DNA concentration in each plug is based on the assumption that each staphylococcal cell contains 0.3×10^{-14} g of DNA since the *S. aureus* chromosome is approximately 3 Mb in size. These are only estimations of DNA concentration, and may be subject to error in the relationship between A_{600} and cells mL^{-1}. Each organism should be assessed for the relationship between cell number and amount of DNA per plug.

The important final criterion to apply to DNA concentration in the plug is that the amount of DNA per band on the gel is optimal, and the amounts per gel lane are equal or nearly so, to allow accurate measurement or comparison of migration distances and numbers of bands in the gel lanes.

5.1.3 Embedding and Lysing Cells in Agarose

The proper design of an insert mould is important for the production of good agarose plugs and thus good quality PFGE gels. The dimensions of the mould should be designed to fit the internal dimensions of the teeth of the electrophoresis comb used. This ensures that the agarose plugs will fit closely but readily into the wells of the electrophoresis gel. The mould block should be made of perspex or other hydrophobic plastic. Moulds which produce square agarose plugs are better than ones which produce rectangular plugs (e.g. the mould used in our laboratory is 10 mm × 10 mm × 1.5 mm), because square plugs are less prone to damage than rectangles; long thin agarose plugs break easily. If plug release is difficult, the mould may be sprayed before use with a commercial mould releaser.

The cell suspension is warmed to 37°C, mixed with an equal volume of 1% low melting temperature agarose in TE buffer at 48°C, and distributed into the mould (10 × 10 × 1.5 mm, or 150 µL per plug) which has been sealed at the bottom with a strip of plastic tape. The mould block is cooled on ice for 10 min, then using a perspex piece which is a close fit to the mould, the plugs are pushed out into lysis solution in a sterile tube (one mL of lysis solution per plug).

The plugs are incubated at 37°C for 4–5 h in a water bath in which time the plugs pass from opaque to transparent, indicating digestion of cell wall and membranes. When preparing DNA from staphylococci other than S. aureus, overnight incubation in lysing solution may be necessary to obtain complete clearing of the plugs at the enzyme concentrations indicated. However, the lysis step can be excluded when preparing DNA from gram negative bacteria [26], in which case the agarose plugs are treated directly with proteinase K as described below.

The lysis solution is discarded and the plugs are rinsed with 1–3 mL of ES, then incubated with ESP (1 mL per plug) at 50°C for 24 h. ESP is replaced with TE and the plugs are washed for 2 h at room temperature with (slow) agitation before changing into fresh TE and storing at 4°C. DNA in agarose plugs can be kept for 2 years or more at 4°C if the plugs are carefully prepared.

DNA for PFGE can also be prepared in an agarose matrix in the form of microbeads [27].

5.1.4   Restriction Endonuclease Digestion

Plugs are sliced into half plugs (i.e. 10 mm × 5 mm × 1.5 mm) containing approx. 1 µg of DNA using a sterile surgical blade and a small spatula (sterilized by dipping into absolute ethanol and flaming) to constrain the plug. The plugs are then washed successively in TE (4–6 changes over 2 h) and the appropriate restriction enzyme buffer (without bovine serum albumin (BSA) or enzyme, 6 changes over 2 h), all at room temperature. The washes are carried out in sterile 60 mm tissue culture dishes on a platform rocker at slow speed, using

5 mL of wash solution per half plug.

Half plugs are then transferred into Eppendorf tubes and incubated in 250 μL of appropriate restriction buffer containing nuclease-free BSA (200 μg mL⁻¹) and 10 U endonuclease. Digestion is carried out overnight, without agitation, at the temperature recommended for that endonuclease. After digestion, the endonuclease buffer is replaced with ES. It is not necessary to treat the plugs to remove or inactivate restriction endonucleases. Digested DNA in plugs is stable in ES at 4°C for months provided care is taken to avoid contamination with DNAases.

5.1.5   Electrophoretic Procedures

(a) Gel preparation and loading of plugs: Electrophoresis gels (normally 1% type II agarose, medium EEO, Sigma or equivalent) are poured in sizes according to the number of samples to be tested (100 mL for a casting tray of 14 cm × 12.7 cm; 150 mL for a 20.9 cm × 12.7 cm; or 225 mL for a 20 cm × 20 cm tray). Varying agarose concentration beyond the range 0.8–1.2% has little effect (see section 4 above). The agarose is dissolved in 0.5 × TBE by heating either in a boiling water bath or directly in a microwave oven (2 min at high power per 100 mL) until no particles of agarose are evident in the molten gel. When using a microwave oven to dissolve the agarose, water lost by evaporation should be replaced before pouring the gel. In a 500 mL conical flask loosely covered with a glass petri dish, about 10 mL of water evaporates in 2 min at high power in a microwave oven. The loss can be readily estimated by weighing the flask and its contents before and after heating. Adding cold water to make up the loss assists cooling the gel to 60°C before pouring.

TBE (0.5×) is also used in the electrophoresis bath, with continuous recirculation. Other electrophoresis buffers can be used but TBE has the advantage of a higher buffering capacity during extended electrophoresis [28]. λ DNA concatemers are used for fragment size calibration of the gel; these can either be purchased, or made by mixing freshly propagated λ phage particles with low melting temperature agarose in a 1 : 1 ratio, and forming plugs and digesting in ESP as described above for cell lysis and digestion. The λ DNA polymerizes spontaneously, by virtue of the cohesive single-stranded ends of the phage genome, generating an incremental series, or 'ladder', of concatemers up to 20–30 units (1–1.5 Mb) in length. A detailed protocol is given in [29]. λ concatemers can also be prepared directly from λ DNA, provided the DNA has not previously been frozen [8].

Loading the plugs into the agarose wells should be done carefully to ensure that the plugs are not damaged and are positioned against the front wall of the well and parallel to it. This ensures sharp and straight DNA bands after electrophoresis. The plugs are then sealed in with molten low melting temperature agarose (1% in TE). Some workers melt the digested agarose plugs at 65°C and load them in the molten form. The loading of plugs into wells is a skill requiring practice and perseverance.

(b) Electrophoresis: Before the loaded gel is placed in the electrophoresis chamber, the temperature of the chamber buffer should be brought to 12–14°C by running the cooling system and the circulating pump. The loaded electrophoresis gel is then placed flat onto the support surface in the electrophoresis tank ensuring that no air bubbles are trapped under the gel; these will interfere with the even transfer of heat, causing irregular migration of DNA molecules.

The electrophoretic field and pulse parameters depend on the number and size of the DNA fragments to be resolved. Restriction endonucleases that generate between 20–30 fragments with a broad size distribution (say 10–1000 kb) give bands on gels which should be well resolved for the most part. However, the selection of endonucleases is a matter of trial and error guided by the nucleotide composition and approximate size of the genome; it will also be governed by the purpose of the PFGE analysis. Greater numbers of fragments decrease the degree of resolution of bands on PFGE gels, and accordingly will not allow recognition of individual fragments for sizing or mapping, or for unequivocal fragment isolation.

When staphylococcal DNA is cut with *Sma*I, *Csp*I or *Sac*II about 10–25 fragments are generated with a size range of 10–700 kb. This is near optimum for mapping and comparative analysis of genomic similarity. The following parameters produce a good resolution of fragments in this size range, using FIGE or CHEF electrophoresis systems.

FIGE programming:	*CHEF programming:*
Field: 180 V (5.5 V cm⁻¹)	Field: 200 V (6 V cm⁻¹)
First parameter settings:	First parameter settings:
initial forward pulse = 1 s	initial forward pulse = 23 s
final forward pulse = 4 s	final forward pulse = 23 s
ratio (forward to reverse) = 3 : 1	
time = 7 h	time = 4 h

Second parameter settings:
  initial forward pulse = 4 s
  final forward pulse = 30 s
  ratio = 3 : 1
  time = 25 h

Second parameter settings:
  initial forward pulse = 1 s
  final forward pulse = 40 s

  time = 22 h

When electrophoresis is completed, the gels are stained with ethidium bromide ($0.5\,\mu g\,mL^{-1}$ in water for 1 h), destained in water for 2–4 h, and photographed over a UV transilluminator.

Conditions suited to other types of bacteria are referred to in Section 7 below.

5.1.6   Troubleshooting

Symptom	Possible Cause	Remedy
Smearing of DNA along gel lanes	Presence of non-specific DNAases; incorrect pulse time or field strength; overloading with DNA	Sterilize all solutions and equipment; test heat-labile reagents for DNAases; use gloves; alter field and/or pulse conditions
DNA remains in well or appears as single high MW band	Undigested DNA due to inactive endonuclease	Wash plugs more extensively before digestion to remove detergents and proteinase; use correct restriction buffer; use active endonuclease
Faint bands amongst strong bands	Partial digestion due to low endonuclease activity; incorrect or inhibitory digestion condition	Check endonuclease and BSA concentration; check buffer and temperature; check activity of enzyme
DNA bands not straight across gel ('smiling')	Inadequate or uneven cooling of running gel by electrophoresis buffer	Check rate of buffer circulation and buffer temperature
Curving of outside lanes (inner lanes straight)	High buffer flow lifting gel away from support	Decrease buffer flowrate
Distorted (ragged) bands	Broken or distorted plug	Handle plug with care; ensure plug is aligned against front of well before sealing into electrophoresis gel

Symptom	Possible Cause	Remedy
Lane trajectories	Erratic heat exchange due to trapped air bubbles beneath electrophoresis gel	Ensure gel does not trap air when loaded onto support plate in chamber; keep circulating buffer volume minimal
Fuzzy (thick) bands	Running gel too thick; distortion of plugs when wells are sealed	Reduce electrophoresis gel thickness; control temperature of LMT agarose used for sealing plug into gel
Increased migration rate (at same pulse and field parameters), high current flow	Excess salt in buffer	Check composition and pH of buffer

5.2   Further Analysis of DNA Fragments

5.2.1   Recovery of Fragments

Large fragments of DNA can be recovered from PFGE gels for further analysis or processing provided that high purity agarose (such as Sea Plaque low melting-temperature agarose) is used in the running gel. DNA fragments in blocks excised from the first electrophoresis gel can be digested with further endonucleases followed by another electrophoretic fractionation (two dimensional electrophoresis ([30]; see also Chapters 12 and 13), or the DNA extracted and used for cloning into recombinant vectors. DNA fragments recovered from PFGE gels can also be used as probes to identify homology between genomic fragments for genome mapping studies. Kits are available for random priming of DNA in LMT agarose (e.g. Megaprime DNA labelling system, Amersham) [31]. DNA fragments can also be purified from LMT agarose by phenol extraction and ethanol precipitation [26], and labelled by nick-translation [28].

5.2.2   Transfer of DNA to membrane filters

For the transfer of DNA fragments in PFGE gels to nitrocellulose filters, the standard method of Southern [32] is used with the following modifications to facilitate the transfer of high molecular weight DNA from gel to filter:

(i) DNA in gels is depurinated by soaking twice in 0.25 M HCl for 15 min;

(ii) transfer is carried out for 24 h or more.

When transferring DNA from PFGE gels to nylon membranes, the method of alkaline Southern blotting described by Reed [33] is used, except that transfer is done in 10 mM NaOH for 24 h instead of 0.4 M NaOH for 4–6 h (Matthaei, personal communication).

6. PFGE ANALYSIS OF BACTERIAL GENOMES

6.1 Genotyping

Analysis of large genomic fragments produced by restriction endonuclease digestion of DNA and separated by PFGE provides a precise and convenient method of genotyping bacteria. Because the distribution of restriction endonuclease sites in a DNA molecule is a function of the nucleotide sequence of the molecule, comparison of the similarity in size and number of fragments generated by restriction endonuclease digestion of genomic DNA gives a useful approximation of the degree of sequence similarity between strains. The precision of the similarity measure increases with the total number of fragments examined; in practice, the upper limit on the number of fragments is the ability to distinguish between individual bands on an electrophoresis gel. These differences in fragment size and number are referred to as *restriction fragment length polymorphisms* (RFLPs). The resolution limit can be extended by examining RFLP patterns generated by more than one enzyme, in separate or combined digestions resolved on a single gel, or by two dimensional PFGE (Chapter 13).

Fragments resolved on PFGE after digestion can be compared for putative identity by determining whether they migrate at precisely the same speed (by eye, or by densitometry of bands on photographic negatives obtained after staining the DNA with ethidium bromide). Values of F (fraction of 'identical' fragments [34]), or %S (the percentage of similarity or Dice coefficient [35]) can then be calculated, and from these a statistical estimate of p (sequence similarity, as the fraction of nucleotides that are the same [36]) can be obtained. Values of F or p can be used to generate similarity matrices; from these, dendrograms can be constructed, giving a visual representation of similarity relationships among numbers of isolates. A number of computational programs are available, for example TOPOL software [37], or simpler software for use in microcomputers and personal computers (see, for example, MacClade [38], TreeDraw [39]). This software may be purchased from computer companies, or often is accessible free or at low cost from networks specializing in biological data storage and processing.

The clustering of isolates in dendrograms can be used as the basis for assigning genetic or nomenclatural divisions in populations of bacteria.

Genotyping using PFGE compares well with other methods of typing. It provides higher resolution of subgroups compared with methods based on phage susceptibility [40], antibiotic resistance [40], serology [41] or carriage of plasmids [42]. This is expected, as PFGE of endonuclease-generated fragments is capable of sampling a greater fraction of the genome for differences. PFGE typing rivals multilocus enzyme electrophoretic typing [43], which also measures differences arising from a significant proportion of the genome, and ribosomal RNA gene characterization, which samples a limited part of the genome [44].

PFGE genotyping is valuable in epidemiological studies, where precise identification of lineages and subtypes is required. A visual inspection of RFLP patterns can be used to establish qualitatively the identity or otherwise of diverse isolates [45]. Quantitative estimates of these differences can be obtained using numerical analysis of the sort described earlier [40].

RFLP analysis provides information on population diversity. For example, a wide range of fragment patterns is seen amongst *S. aureus* types carried by healthy individuals, indicating genetic diversity amongst the strains. In contrast, RFLP analysis of methicillin-resistant *S. aureus* isolates from widely separated geographic locations indicates little genetic variation amongst the isolates, suggesting that members of this clinically significant subgroup of staphylococci may belong to a single clone [46].

6.2 Sizing, Mapping and Structural Analysis of the Bacterial Genome

The size of a bacterial genome can be estimated by summing the DNA fragments formed by restriction endonuclease digestion of the total genome and resolved by PFGE in parallel with DNA size markers [47]. Small genomes can be measured without endonuclease digestion after linearization of the genome, for example by X-irradiation of the cells before lysing [13].

Restriction endonuclease sites in a chromosome can be mapped using two dimensional PFGE [30], and by cross-hybridizing fragments from one endonuclease digestion with fragments generated by other endonucleases [48]. This large-scale mapping can be followed by finer mapping using specific cloned genes to probe individual PFGE fragments [49]. If suitable restriction endonuclease sites are not present on the genome, a polylinker containing a range of sites for rare-cutting restriction endonucleases can be introduced into a transposon and thence into the chromosome [50]. Alternately, rare-cutting sites can be introduced around the genome using a plasmid [51] or a transposon inserting at multiple sites [52].

Deletions and insertions as small as 2% of the DNA fragment in which they occur can be detected and mapped [20]. The number and location of insertion sequences and transposons in a particular genome can also be determined by hybridization probing of DNA fragments generated by PFGE [53], using probes specific for the mobile elements.

The higher level structure of bacterial DNA can be studied using PFGE. Large circular DNA molecules migrate much more slowly in PFGE than do the corresponding linear forms. This phenomenon has been used to detect linearity of chromosomal DNA, where the genome is small. Thus, the *Borrelia* chromosome appears to be linear in the natural state [54], in contrast to that of the related *Leptospira* and most other bacterial chromosomes, which are circular. The anomalous behaviour of plasmids in PFGE is also a consequence of the different migratory rates of circular and linear molecules and should be borne in mind when interpreting the PFGE patterns of total genomic DNA.

Genome mapping is discussed further in Chapter 12.

7. SURVEY OF BACTERIAL DNA EXAMINED BY PFGE

Organism (Genome size)	PFGE specifications[a]	Ref.	Enzyme[b]
Acinetobacter	CHEF: 10 s, 21 h; 8 s, 12 h; 150 V	55	Apal, Smal
Bacillus (4.7 Mb)	CHEF: 20 s, 40 h; 90 V; 13°C	56	NotI
Borrelia (946 kb)	CHEF: 3 s, 9.5 h; 5 s, 8.5 h; 25 s, 4 h	54	Mlul, Smal, Saccl, BssHI
Bradyrhizobium	FIGE: 0.3 s–28.5 s forward, ratio 3 : 1; 20 h; 10 V cm⁻¹; 13°C	57	Asel, Spel
Brucella (3.1 Mb)	CHEF: 20–5 s; 40 h; 200 V	48	Spel, Xhol, Xbal
Campylobacter (1.7 Mb)	FIGE: 0.1–10 s forward, ratio 2 : 1, 4 h ×6; then 0.1–55 s forward, ratio 2 : 1, 4 h × 6; RT	58	BssHI, Ncil, Sall, Smal
Caulobacter (3.8 Mb)	OFAGE: 3 to 30 s; 14 to 16 h; 230 V	59	Dral, Asel, Spel
Chlamydia (1.4 Mb)	CHEF: 200 s, 100 s, 60 s, 30 s; 12 h; 150 V	60	NotI, Sfil
Clostridium (3.58 Mb)	FIGE: 0.33-60 s forward, ratio 3 : 1; 36 h; 240 V; 8°C	61	Apnl, Fspl, Mlnl,Nrul, Saccll, Smal
Enterococcus	CHEF: 5-35 s; 30 h; 200 V	62	Smal, NotI
Escherichia	OFAGE: 20 s; 72 h; 250 V; 14°C	63	Xbal, NotI
	CHEF: 10 to 40 s; 25 h; 200 V; 14°C	64	
Haemophilus (2 Mb)	FIGE: 0.03–9 s forward, ratio 3 : 1 (fr. <200 kb); 0.3–30 s forward, ratio 3 : 1(fr. >200 kb), 16 h; 240 V; 8°C	47	Smal, Apal, NaelRsrll, Eagl
	OFAGE: 1–36 s; 12 h; 280 V	25	
	CHEF: 20 to 30 s; 40 h; 8.3 V cm⁻¹; 6°C	26	
Lactococcus (2.6 Mb)	CHEF: 1–70 s; 22 h; 6 V cm⁻¹; 15°C	65	NotI, Smal, Apal
Legionella	CHEF: 7–74 s; 24 h; 200 V; 16°C	44	Sfil, NotI
Leptospira (3.1–4.4 Mb)	CHEF: 5–50 s, 22 h; 70–90 s, 2 h (fr. <750 kb); 10–70 s, 24 h (fr. >750 kb); 200 V; 10°C	66	NotI, Smal, Apal,Sfil, SgrAl
Listeria (2.6 Mb)	CHEF: 25–5 s; 43 h; 180 V FIGE: 1.5–18 s forward, ratio 3 : 1; 22 h; 9°C	67	Apal, NotI, Smal
Mycobacterium	CHEF: 5–20 s; 20 h; 200 V; 14°C	45	Xbal, Dral, Spel Asnl
	TAFE: 20–60 s; 18–20 h; 150 mA; 12°C	45	
	FIGE: 0.33–60 s forward, ratio 3 : 1; 36 h; 100 V; 18°C	68	

Mycoplasma (0.8–1 Mb)	CHEF: 1–15 s; 20 h; 200 V; 14°C	49	Smal, Apal,
	FIGE: 0.45–28.8 s forward, ratio 3 : 1; 6 h; 200 V; 15°C	69	Sfil,Nrul, BamHI
Myxococcus (9.5 Mb)	CHEF: 70 s, 15 h; 120 s, 12 h; 170 V; 14°C	70	Asel
	FIGE: 2.4 s initial forward, ratio 3 : 1, ramp factor 2 h⁻¹; 8 h; 135 V; 25°C	70	
Neisseria (2.2 Mb)	CHEF: 12 s (fr. <250 kb), 60 s (fr. >250 kb); 40–72 h; 100 V; 15°C	71	Nhel, Spel
Porochlamydia (2.6, 1.4 Mb)	CHEF: 200 s, 100 s, 60 s, 30 s; 12 h; 150 V	60	
Pseudomonas	CHEF: 5 s, 24 h; 10 s, 24 h; 200 V	72	Smal, Spel
	TAFE: 5 s, 1 h; 30 s, 18 h ×1–3; 240 V	73	Xbal,Sspl, Nhel
	FIGE: 96 h, ratio 3 : 1; 5 V cm⁻¹; 10°C	74	Dral
Rickettsiella (2.1, 1.7 Mb)	CHEF: 60 s, 24 h; 30 s; 24 h; 150 V	60	Notl, Sfil
Salmonella (4.8 Mb)	TAFE: 120 s, 36 h	52	Blnl, Spel, Xbal
	CHEF: 5–40 s; 24 h; 200 V	75	Smal, Pacl
Shigella	FIGE: 25–29 h; 180–190 V; 14°C	76	Notl
Staphylococcus (3.1 Mb)	CHEF: 23 s; 4 h; 1–40 s, 22 h; 200 V; 14°C	46	Smal, Saccll Eagl, Sstll
	TAFE: 4 s, 1 h; 8 s, 10 h; 20 s, 3 h; 8 s, 4 h; 150 mA; 13°C	77	Notl
	FIGE: 1.2–12 s forward, 3 h; 0.75 s forward, 0.5 h, ratio 3 : 1 (fr. >50 kb); 0.4 s forward, 3.25 h, ratio 2 : 1 (fr. <50 kb); 16 V cm⁻¹; 18°C	78	
Stigmatella (Myxobacteria) (9.35 Mb)	CHEF: 30–300 s; 45 h; 150 V; 15°C (fr. 2 Mb)	79	Asel, Spel
Streptococcus (2.2Mb)	FIGE: 0.3–60 s forward, ratio 3 : 1 (fr. >150 kb); 0.3–30 s forward, ratio 3 : 1 (fr. <150 kb); 16 h; 240 V; 8°C	80	Smal, Apal, Saccl, Notl, Sfil
	OFAGE: 2 to 60 s; 24 to 40 h; 330 V; 10°C	81	
	CHEF: 10 to 60 s; 40 h; 170 V; 10°C	81	
Streptomyces (8 Mb)	CHEF: 30 s, 22 h; 80 s, 11 h; 20 s, 9 h; 170 V	82	Asel, Dral, Sspl
	PHOGE: 40 s, 16 h, 200 V; 100 s, 12 h, 170 V; 120 s, 10 h, 170 V	82	
Sulfolobus (3 Mb)	OFAGE: 600 s; 120 h; 75 V; 16°C	83	Notl
Taylorella (1.5 Mb)	FIGE: 6–12 cycles, 0.05–55 s, ratio 3 : 1; 4 h; 90 V	84	Apal, Nael

(Continued over)

Organism (Genome size)	PFGE specifications[a]	Ref.	Enzyme[b]
Thermococcus (1.89 Mb)	TAFE: 4 s, 0.5 h, 330 V; 60 s, 18 h, 250 V (fr. >300 kb); 15 s, 18 h, 150 V (fr. <300 kb); 13°C	85	Nhel, Spel, Xbal
Treponema (900 kb)	CHEF: 1–10 s; 12 h; 200 V; 15°C	86	Notl, Spel
Yersinia (4.7 Mb)	FIGE: 3.75–11.25 s forward, ratio 3 : 1; 24 h; 200 V; 10°C	87	Notl, Sfil
	CHEF: 1 to 20 s; 200 V; 10°C	87	

Terminology: 1–3 s, ramp from 1 s to 3 s; 1 to 3 s, fixed pulse time, chosen from the range 1–3 s.
Abbrev.:   fr., fragment; RT, room temperature.
[a]Pulse time (seconds); runtime (hours); voltage; and temperature are listed; where omitted, these details were not provided in reference.
[b]Enzymes used in all references are listed.

REFERENCES

[1]   Schwartz, D.C., Safran, W., Welch, J., Haas, J., Goldenburg R.M. and Cantor, C.R. New techniques for purifying large DNAs and studying their properties and packaging. Cold Spr. Harb. Symp., 47 (1983) 189–195.

[2]   Smith, C.L., Lawrance, S.K., Gillespie, G.A., Cantor, C.R., Weissman, S.M. and Collins, F.S. Strategies for mapping and cloning macroregions of mammalian genomes. Methods Enzymol., 151 (1987) 461–489.

[3]   Krawiec, S. and Riley, M. Organization of the bacterial chromosome. Microbiol. Rev., 54 (1990) 502–539.

[4]   Carle, G.F., Frank, M. and Olson, M.V. Electrophoretic separation of large DNA molecules by periodic inversion of the electric field. Science, 232 (1986) 65–68.

[5]   Schwartz, D.C. and Cantor, C.R. Separation of yeast chromosome-sized DNAs by pulsed field gradient gel electrophoresis. Cell, 37 (1984) 67–75.

[6]   Carle, G.F. and Olson, M.V. Separation of chromosomal DNA molecules from yeast by orthogonal-field-alternation gel electrophoresis. Nucleic Acids Res., 12 (1984) 5647–5664.

[7]   Chu, G., Vollrath, D. and Davis, R.W. Separation of large DNA molecules by contour-clamped homogeneous electric fields. Science, 234 (1986) 1582–1585.

[8]   Bancroft, I. and Wolk, C.P. Pulsed homogeneous orthogonal field gel electrophoresis (PHOGE). Nucleic Acids Res., 16 (1988) 7405–7418.

[9]   Clark, S.M., Lai, E., Birren, B.W. and Hood, L. A novel instrument for separating large DNA molecules with pulsed homogeneous electric fields. Science, 241 (1988) 1203–1205.

[10] Ziegler, A., Geiger, K.H., Ragoussis, J. and Szalay, G. A new electrophoresis apparatus for separating very large DNA molecules. J. Clin. Chem. Clin. Biochem., 25 (1987) 578–579.

[11] Gardiner, K. and Patterson, D. Transverse alternating field electrophoresis and

applications to mammalian genome mapping. Electrophoresis J., 10 (1989) 296–301.

[12] Turmel, C., Brassard, E., Forsyth, R., Hood, K., Slater, G.W. and Noolandi, J. High resolution zero integrated field electrophoresis (ZIFE) of DNA. In: Birren, B. and Lai, E. (eds.), *Electrophoresis of Large DNA Molecules*, Cold Spring Harbor Laboratory, Cold Spring Harbor, N.Y., 1990, pp.101–131.

[13] Gunderson, K. and Chu, G. Pulsed–field electrophoresis of megabase-sized DNA. Mol. Cell. Biol., 11 (1991) 3348–3354.

[14] Stellwagen, N. C. Effect of the electric field on the apparent mobility of large DNA fragments in agarose gels. Biopolymers, 24 (1985) 2243–2255.

[15] Deutsch, J.M. Theoretical studies of DNA during gel electrophoresis. Science, 240 (1988) 922–924.

[16] Deutsch, J.M. and Madden, T.L. Theoretical studies of DNA during gel electrophoresis. J. Chem. Phys., 94 (1989) 2476–2485.

[17] Madden, T.L. and Deutsch, J.M. Theoretical study of DNA during orthogonal field alternating gel electrophoresis. J. Chem. Phys., 94 (1991) 1584–1591.

[18] Turmel, C., Brassard, E., Slater, G.W. and Noolandi, J. Molecular detrapping and band narrowing with high frequency modulation of pulsed field electrophoresis. Nucleic Acids Res., 18 (1990) 569–575.

[19] Smith, C.L., Warburton, P.E., Gaal, A. and Cantor, C.R. Analysis of genome organization and rearrangements by pulsed field gel gradient gel electrophoresis. Genet. Eng., 8 (1986) 45–70.

[20] Inglis, B., Matthews, P.R. and Stewart, P.R. Induced deletions within a cluster of resistance genes in the mec region of the chromosome of *Staphylococcus aureus*. J. Gen. Microbiol., 136 (1990) 2231–2239.

[21] Inglis, B., El-Adhami, W. and Stewart, P.R. Methicillin-sensitive and -resistant homologues of *Staphylococcus aureus* occur together among clinical isolates. J. Infect. Dis., 167 (1993) 323–328.

[22] Stewart, P.R., El-Adhami, W., Inglis, B. and Franklin, J.C. Analysis of an outbreak of variably methicillin-resistant *Staphylococcus aureus* with chromosomal RFLPs and mec region probes. J. Med. Microbiol., 38 (1993) 270–277.

[23] Ichiyama, S., Ohta, M., Shimokata, K., Kato, N. and Takeuchi, J. Genomic DNA fingerprinting by pulsed-field gel electrophoresis as an epidemiological marker for study of nosocomial infections caused by methicillin–resistant *Staphylococcus aureus*. J. Clin. Microbiol., 29 (1991) 2690–2695.

[24] Poddar, S.K. and McClelland, M. Restriction fragment fingerprinting and genome sizes of Staphylococcus species using pulsed-field gel electrophoresis and infrequent cleaving enzymes. DNA Cell Biol., 10 (1991) 663–669.

[25] Lee, J.J. and Smith, H.O. Sizing of the Haemophilus influenzae Rd genome by pulsed-field genome electrophoresis. J. Bacteriol., 170 (1988) 4402–4405.

[26] Butler, P.D. and Moxon, E.R. A physical map of the genome of Haemophilus influenzae type b. J. Gen. Microbiol., 136 (1990) 2333–2342.

[27] Carle, G.F. and Olson, M.V. Orthogonal-field-alternation gel electophoresis. Methods Enzymol., 155 (1987) 468–482.

[28] Maniatis, T., Fritsch, E.F. and Sambrook, J. *Molecular Cloning: a Laboratory Manual.* Cold Spring Harbor Laboratory, Cold Spring Harbor, N.Y., 1982, p.156, pp. 109–112.

[29] Waterbury, P.G. and Lane, M.J. Generation of lambda phage concatemers for use as pulsed field electrophoresis size markers. Nucleic Acids Res., 15 (1987) 3930.

[30] Bautsch, W. Rapid physical mapping of the *Mycoplasma mobia* genome by two-dimensional field inversion gel electrophoresis techniques. Nucleic Acids Res., 16 (1988) 11461–11467.

[31] Feinberg, A.P. and Vogelstein, B. A technique for radiolabelling DNA restriction endonuclease fragments to high specific activity. Addendum. Anal. Biochem., 137 (1984) 266–267.

[32] Southern, E.M. Detection of specific sequences among DNA fragments separated by gel electrophoresis. J. Mol. Biol., 98 (1975) 503–517.

[33] Reed, K.C. Nucleic acid hybridizations with positive charge-modified nylon membrane. In: Dale, J.W. and Sanders, P.G. (eds.), *Methods in Gene Technology*, Vol.1, JAI Press, London, 1991, pp. 127–160.

[34] Nei, M. and Li, W.H. Mathematical model for studying genetic variation in terms of restriction endonucleases. Proc. Natl. Acad. Sci. USA, 76 (1979) 5269–5273.

[35] Dice, L.R. Measures of the amount of ecological association between species. Ecology, 26 (1945) 297–302.

[36] Upholt, W.B. Estimation of DNA sequence divergence from comparison of restriction endonuclease digests. Nucleic Acids Res., 4 (1977) 1257–1265.

[37] Lalouel, J. Topology of population structure. In: Morton, N.E. (ed.), *Genetic Structure of Populations*, University of Hawaii Press, Honolulu, 1973, pp. 139–152.

[38] Maddison, W. and Maddison D. Program available by anonymous ftp from EMBL molecular biology software server.

[39] Felsenstein, J. PHYLIP (Phylogeny Interface Package) v.3.3. For information, send electronic mail to Joe@genetics.washington.edu.

[40] Streulens, M.J., Deplano, A., Godard, C., Maes, N. and Serrys, E. Epidemiologic typing and delineation of genetic relatedness of methicillin-resistant *Staphylococcus aureus* by macrorestriction analysis of genomic DNA by using pulsed-field gel electrophoresis. J. Clin. Microbiol., 30 (1992) 2599–2605.

[41] Gordillo, M.E., Reeve, G.R., Pappas, J., Mathewson, J.J., DuPont, H.L. and Murray, B.E. Molecular characterization of strains of enteroinvasive *Escherichia coli* O143, including isolates from a large outbreak in Houston, Texas. J. Clin. Microbiol., 30 (1992) 889–893.

[42] Parisi, J. Coagulase-negative staphylococci and the epidemiological typing of *Staphylococcus epidermidis*. Microbiol. Rev., 49 (1985) 126–139.

[43] Arbeit, R.D., Arthur, M., Dunn, R., Kim, C., Selander, R. and Goldstein, R. Resolution of recent evolutionary divergence amongst *Escherichia coli* from related lineages: the application of pulsed-field electrophoresis to molecular epidemiology. J.Inf.Dis., 161 (1990) 230–235.

[44] Schoonmaker, D., Heimberger, T. and Birkhead, G. Comparison of ribotyping and restriction enzyme analysis using pulsed-field gel electrophoresis for distinguishing *Legionella pneumophila* isolates obtained during a nosocomial outbreak. J. Clin. Microbiol., 30 (1992) 1491–1498.

[45] Hector, J.S., Pang, Y., Mazurek, G.H., Zhang, Y., Brown, B.A. and Wallace, R.J. Jr. Large restriction fragment patterns of genomic *Mycobacterium fortuitum* DNA as strain-specific markers and their use in epidemiologic investigation of four nosocomial outbreaks. J. Clin. Microbiol., 30 (1992) 1250–1255.

[46] El-Adhami, W., Roberts, L., Vickery, A., Inglis, B., Gibbs, A. and Stewart, P.R. Epidemiological analysis of a methicillin–resistant *Staphylococcus aureus*

outbreak using restriction fragment length polymorphisms of genomic DNA. J. Gen. Microbiol., 137 (1991) 2713–2720.

[47] Kauc, L., Mitchell, M. and Goodgal, S.H. Size and physical map of the chromosome of *Haemophilus influenzae*. J. Bacteriol., 171 (1989) 2474–2479.

[48] Allardet-Servent, A., Carles-Nurit, M.J., Bourg, G., Michaux, S. and Ramuz, M. Physical map of the *Brucella melitensis* 16 M chromosome. J. Bacteriol., 173 (1991) 2219–2224.

[49] Ladefoged, S.A. and Christiansen, G. Physical and genetic mapping of the genomes of five *Mycoplasma hominis* strains by pulsed-field gel electrophoresis. J. Bacteriol., 174 (1992) 2199–2207.

[50] Wong, K.K. and McClelland, M. Dissection of the *Salmonella typhimurium* genome by use of a Tn5 derivative carrying rare restriction sites. J. Bacteriol., 174 (1992) 3807.

[51] Le Bourgeois, P., Lautier, M., Mata, M. and Ritzenthaler, P. New tools for the physical and genetic mapping of *Lactococcus* strains. Gene, 111 (1992) 109–114.

[52] Wong, K.K. and McClelland, M. A BlnI restriction map of the *Salmonella typhimurium* LT2 genome. J. Bacteriol., 174 (1992) 1656–1661.

[53] Mitchell, D. and Smit, J. Identification of genes affecting production of the adhesion organelle of *Caulobacter crescentus* CB2. J. Bacteriol., 172 (1990) 5425–5431.

[54] Davidson, B.E., MacDougall, J. and Saint-Girons, I. Physical map of the linear chromosome of the bacterium *Borrelia burgdorferi* 212, a causative agent of Lyme disease, and localization of rRNA genes. J. Bacteriol., 174 (1992) 3766–3774.

[55] Gouby, A., Carles-Nurit, M.J., Bouziges, N., Bourg, G., Mesnard, R. and Bouvet, P.J. Use of pulsed-field gel electrophoresis for investigation of hospital outbreaks of *Actinetobacter baumannii*. J. Clin. Microbiol., 30 (1992) 1588–1591.

[56] Ventra, L. and Weiss, A.S. Transposon-mediated restriction mapping of the *Bacillus subtilis* chromosome. Gene, 78 (1989) 29–36.

[57] Sobral, B.W., Sadowsky, M.J. and Atherly, A.G. Genome analysis of *Bradyrhizobium japonicum* serocluster 123 field isolates by using field inversion gel electrophoresis. Appl. Environ. Microbiol., 56 (1990) 1949–1953.

[58] Nuijten, P.J., Bartels, C., Bleumink-Pluym, N.M., Gaastra, W. and van der Zeijst, B.A. Size and physical map of the *Campylobacter jejuni* chromosome. Nucleic Acids Res., 18 (1990) 6211–6214.

[59] Ely, B and Gerardot, C.J. Use of pulsed-field-gradient gel electrophoresis to construct a physical map of the *Caulobacter crescentus* genome. Gene, 68 (1988) 323–333.

[60] Frutos, R., Pages, M., Bellis, M., Roizes, G. and Bergoin, M. Pulsed-field gel electrophoresis determination of the genome size of obligate intracellular bacteria belonging to the genera *Chlamydia*, *Rickettsiella* and *Porochlamydia*. J. Bacteriol., 171 (1989) 4511–4513.

[61] Canard, B. and Cole, S.T. Genome organization of the anaerobic pathogen *Clostridium perfringens*. Proc. Natl. Acad. Sci. U.S.A., 86 (1989) 6676–6680.

[62] Murray, B.E., Singh, K.V., Heath, J.D., Sharma, B.R. and Weinstock, G.M. Comparison of genomic DNAs of different enterococcal isolates using restriction endonucleases with infrequent recognition sites. J. Clin. Microbiol., 28 (1990) 2059–2063.

[63] Ott, M., Bender, L., Blum, G., Schmittroth, M., Achtman, M., Tschape, H. and

Hacker, J. Virulence patterns and long-range genetic mapping of extraintestinal *Escherichia coli* K1, K5 and K100 isolates: use of pulsed-field gel electrophoresis. Infect. Immun., 59 (1991) 2664–2672.

[64] Zingler, G., Ott, M., Blum, G., Falkenhagen, U., Naumann, G., Sokolowska-Kohler, W. and Hacker, J. Clonal analysis of *Escherichia coli* serotype O6 strains from urinary tract infections. Microb. Pathog., 12 (1992) 299–310.

[65] Tulloch, D.L., Finch, L.R., Hillier, A.J. and Davidson, B.E. Physical map of the chromosome of *Lactococcus lactis* subsp. lactis DL11 and location of six putative rRNA operons. J. Bacteriol., 173 (1991) 2768–2775.

[66] Zuerner, R.L. Physical map of chromosomal and plasmid DNA comprising the genome of *Leptospira interrogans*. Nucleic Acids Res., 19 (1991) 4857–4860.

[67] Carriere, C., Allardet-Servent, A., Bourg, G., Audurier, A. and Ramuz, M. DNA polymorphism in strains of *Listeria monocytogenes*. J. Clin. Microbiol., 29 (1991) 1351–1355.

[68] Levy-Frebault, V.V., Thorel, M.F., Varnerot, A., and Gicquel, B. DNA polymorphism in *Mycobacterium paratuberculosis*, 'wood pigeon mycobacteria', and related mycobacteria analysed by field inversion gel electrophoresis. J. Clin. Microbiol., 27 (1989) 2823–2826.

[69] Krause, D.C. and Mawn, C.B. Physical analysis and mapping of the *Mycoplasma pneumoniae* chromosome. J. Bacteriol., 172 (1990) 4790–4797.

[70] Chen, H.W., Kuspa, A., Keseler, I.M.and Shimkets, L.J. Physical map of the *Myxococcus xanthus* chromosome. J. Bacteriol., 173 (1991) 2109–2115.

[71] Dempsey, J.A., Litaker, W., Madhure, A., Snodgrass, T.L. and Cannon, J.G. Physical map of the chromosome of *Neisseria gonorrhoeae* FA1090 with locations of genetic markers, including opa and piuI genes. J. Bacteriol., 173 (1991) 5476–5486.

[72] Allardet-Servent, A., Bouziges, N., Carles-Nurit, M.J., Bourg, G., Gouby, A. and Ramuz, M. Use of low-frequency-cleavage restriction endonucleases for DNA analysis in epidemiological investigations of nosocomial bacterial infections. J. Clin. Microbiol., 27 (1989) 2057–2061.

[73] Anderson, D.J., Kuhns, J.S., Vasil, M.L., Gerding, D.N. and Janoff, E.N. DNA fingerprinting by pulsed field gel electrophoresis and ribotyping to distinguish *Pseudomonas cepacia* isolates from a nosocomial outbreak. J. Clin. Microbiol., 29 (1991) 648–649.

[74] Boukadida, J., de Montalembert, M., Gaillard, J.L., Gobin, J., Grimont, F., Girault, D., Veron, M. and Berche, P. Outbreak of gut colonization by *Pseudomonas aeruginosa* in immunocompromised children undergoing total digestive decontamination: analysis by pulsed-field electrophoresis. J. Clin. Microbiol., 29 (1991) 2068–2071.

[75] Pang, T. Personal Communication.

[76] Soldati, L. and Piffaretti, J.C. Molecular typing of *Shigella* strains using pulsed field gel electrophoresis and genome hybridization with insertion sequences. Res. Microbiol., 142 (1991) 489–498.

[77] Prevost, G., Jaulhac, B. and Piemont, Y. DNA fingerprinting by pulsed-field gel electrophoresis is more effective than ribotyping in distinguishing among methicillin-resistant *Staphylococcus aureus* isolates. J. Clin. Microbiol., 30 (1992) 967–973.

[78] Goering, R.V. and Winters, M.A. Rapid method for epidemiological evaluation of gram-positive cocci by field inversion gel electrophoresis. J. Clin. Microbiol., 30 (1992) 577–580.

[79] Neumann, B., Pospiech, A. and Schaerer, H.U. Size and stability of the

genomes of the Myxobacteria *Stigmatella aurantiaca* and *Stigmatella erecta*. J. Bacteriol., 174 (1992) 6307–6310.

[80] Gasc, A.M., Kauc, L., Barraille, P., Sicard, M. and Goodgal, S. Gene localization, size, and physical map of the chromosome of *Streptococcus pneumoniae*. J. Bacteriol., 173 (1991) 7361–7367.

[81] Okahashi, N., Sasakawa, C., Okada, N., Yamada, M., Yoshikawa, M., Tokuda, M., Takahashi, I. and Koga, T. Construction of NotI restriction map of the *Streptococcus mutans* genome. J. Gen. Microbiol., 136 (1990) 2217–2223.

[82] Kieser, H.M., Kieser, T. and Hopwood, D.A. A combined genetic and physical map of *Streptomyces coelicolor* chromosome. J. Bacteriol., 174 (1992) 5497–5507.

[83] Yamagishi, A. and Oshima, T. Circular chromosomal DNA in the sulfur-dependent archaebacterium *Sulfolobus acidocaldarius*. Nucleic Acids Res., 18 (1990) 1133–1136.

[84] Bleumink-Pluym, N., ter Laak, E.A. and van der Zeijst, B.A. Epidemiologic study of *Taylorella equigenitalis* strains by field inversion gel electrophoresis of genomic restriction endonuclease fragments. J. Clin. Microbiol., 28 (1990) 2012–2016.

[85] Noll, K.M. Chromosome map of the thermophilic archaebacterium *Thermococcus celer*. J. Bacteriol., 171 (1989) 6721–6725.

[86] Walker, E.M., Arnett, J.K., Heath, J.D. and Norris, S.J. *Treponema pallidum* subsp. pallidum has a single, circular chromosome with a size of approximately 900 kilobase pairs. Infect. Immun., 59 (1991) 2476–2479.

[87] Romalde, J.L., Iteman, I. and Carniel, E. Use of pulsed field gel electro-phoresis to size the chromosome of the bacterial fish pathogen *Yersinia rucheri*. FEMS Microbiol. Lett., 68 (1991) 217–225.

[88] Bostock, C. Parameters of field inversion gel electrophoresis for the analysis of pox virus genomes. Nucleic Acids Res. 16 (1988) 4239–4252.

Chapter 12

PULSED-FIELD GEL ELECTROPHORESIS
LARGE SCALE RESTRICTION MAPS

Nobuo Okahashi and Toshihiko Koga

OUTLINE

Methods in Gene Technology, Volume 2, pages 207–226
Copyright © 1994 JAI Press Ltd
All rights of reproduction in any form reserved.
ISBN: 1-55938-264-3

1.   INTRODUCTION

It is essential for understanding of a biological system that the structure and function of genes and their arrangement on a chromosome are elucidated. This includes the physical mapping of the organization of genes on the chromosome by means of ordering of restriction sites at measured distances through chromosomal regions. Information obtained from physical maps will aid in developing new strategies for analysis of regions of high biological importance and for solving genetically medical problems such as malignancy and inherited genetic diseases.

Physical mapping of large DNA molecules initially focused on bacteria [1,2], yeast [3,4] and nematodes [5]. At first, a 'bottom-up' strategy was used to construct physical maps. In this strategy with λ or cosmid clones, large numbers of clones have been analysed by restriction mapping or fingerprinting in order to arrange overlapping clones into contigs through the recognition of shared DNA fragments.

Recent advances in procedures for separating and cloning large DNA fragments allows a 'top-down' strategy for constructing physical maps of genomes. The first step of this strategy is to construct low-resolution restriction maps of genomes. These maps are made by using restriction enzymes that have very infrequent cleavage sites (rare-cutting restriction enzymes). Many investigators have reported low-resolution restriction maps of chromosomes of various bacteria [2,6,7] and human chromosomes [8,9,10,11]. In another 'top-down' strategy, large DNA fragments (up to 1 Mb pairs are cloned in yeast artificial chromosome (YAC) vectors [12, 13]. These vectors can cover comprehensively the target genome with fewer clones. This strategy has made it possible to construct 'cloned maps' of many mammalian chromosomes [14,15,16].

The 'top-down' strategies necessitate the separation of large DNA fragments. Standard procedures such as agarose gel electrophoresis do not permit the resolution of DNA fragments larger than about 50 kb. On the other hand, DNA molecules as large as 10 Mb are resolved by pulsed field gel electrophoresis (PFGE) [3,17,18,19] (see also Chapters 11 and 13). The ability of PFGE to resolve such large fragments will greatly facilitate the construction of physical maps. This method allows the mapping of rare-cutting restriction enzyme sites across large chromosomal regions (e.g. 0.1–10 Mb). The combination of PFGE with mapping techniques may offer a considerable potential for generating restriction map data for significant portions of large genomes such as human genome.

In this chapter, we describe a procedure for constructing a physical map with rare-cutting restriction enzymes. The procedure includes the digestion of chromosomal DNA with rare-cutting restriction enzymes, separation of DNA fragments by PFGE, and Southern blot analysis with linking clones. Recent advances in YAC cloning technique will be described elsewhere.

2. PRINCIPLE

PFGE is a procedure which separates DNA fragments in the range of 50 kb to over 10 Mb; the general principles of PFGE are described in Chapter 11. The development of PFGE has made it possible to determine the size of chromosomal DNAs of several organisms such as yeast and protozoa. However, most mammalian chromosomal DNAs are generally larger than 50 Mb. These large DNAs must be broken into discrete fragments by treating with rare-cutting restriction enzymes for PFGE.

Several 8 bp cutting restriction enzymes are commercially available (Table 1). These enzymes are useful for digestion of mammalian DNAs. A representative 8 bp-cutting restriction enzyme, *Not*I

Table 1.   List of 8 bp restriction enzymes and vectors containing the restriction sites

Enzyme	Recognition site	Vector
*Asc*I	GG CGCGCC	pNEB193
*Fse*I	GGCCGG CC	
*Not*I	GC GGCCGC	pBluescript II, pGEM series (5Zf, 11Zf, 13Zf etc.), λZAP, λGEM-12, λgt11 Sfi-Not
*Pac*I	TTAAT TAA	pNEB193
*Pme*I	GTTT AAAC	pNEB193
*Sfi*I	GGCCNNNN NGGCC	pGEM series (11Zf, 13Zf etc.), λgt11 Sfi-Not
*Srf*I	GCCC GGGC	
*Sse*8387I	CCTGCA GG	pUC series

(GC GGCCGC) cuts the 4.7 Mb genome of *Escherichia coli* K12 into 22 DNA fragments ranging 20 kb to 1 Mb [2]. This restriction enzyme generates 21 [8] or 33 [9] DNA fragments from the human chromosome 21 (50 Mb). DNA fragments generated by the *Not*I digestion can be separated by PFGE. In addition to 8 bp-cutting restriction enzymes, 6 bp-cutting restriction enzymes such as *Mlu*I (A CGCGT) and *Pvu*I (CGAT CG) can be used for generating large fragments from mammalian DNAs [19]. Large DNA fragments generated by digestion with rare-cutting restriction enzymes are separated by PFGE, and transferred to nylon membrane by standard procedures [20].

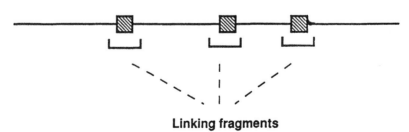

Genomic DNA   Sites for rare-cutting restriction enzyme

Linking fragments

Figure 1.   Schematic representation of linking clones. The linking fragments are segments of chromosomal DNA containing an internal recognition site of a rare-cutting restriction enzyme.

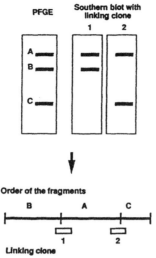

Figure 2. Southern blot analysis with linking clones. The radiolabelled linking clone hybridizes specifically with two adjacent fragments. Linking clones 1 and 2 hybridize to fragments A and B, and fragments A and C, respectively. Therefore, the order of these fragments on the chromosome is B–A–C.

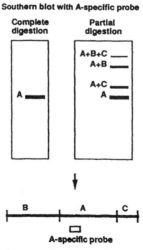

Figure 3. Southern blot analysis with partially digested DNA. When the digestion is complete, an A-specific probe hybridizes only with fragment A. When partial digested DNA is blotted, the probe hybridizes with bands correspond to A+B+C, A+B and A+C. These bands can be identified on the basis of their sizes. From these results, the order of these three fragments is determined to be B–A–C.

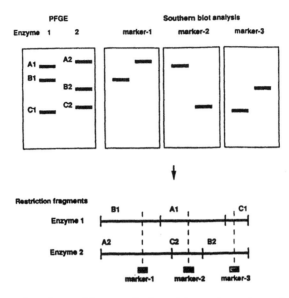

Figure 4. Southern blot analysis with a cloned marker. Southern blot analysis with linking clones shows that the order of fragments A1, B1 and C1 is B1–A1–C1. Southern blot analysis with three radiolabelled markers (markers –1, –2 and –3) shows that the order of fragments A2, B2 and C2 produced by another restriction enzyme (enzyme 2) is A2–C2–B2.

The next step for constructing a physical map is to isolate genomic clones which contain internal sites recognized by rare-cutting restriction enzymes (Figure 1). Clones bearing these fragments are called linking clones [2,19]. Linking clones can be employed to determine the orientation and location of the fragments separated by PFGE on the chromosome. The mapping procedure is summarized in Figure 2. Linking clones are radiolabelled and used as probes. By Southern blot analysis, two adjacent DNA fragments are detected by a linking clone (Figure 2).

There are several strategies for construction of linking libraries [2,6,11,21,22]. Among these strategies, two methods are described in sections 4.1 and 4.2.

As it is difficult to obtain a complete set of linking clones, it is often not possible to construct a physical map by Southern blotting only with a linking library. In such a case, additional procedures are employed to complete a physical map. One method is Southern blotting with partially digested genomic DNA. The principle of this method is shown in Figure 3. In Southern blot analysis with completely digested DNA, an A-specific probe hybridizes only with fragment A (Figure 3, left). On the other hand, the A-specific probe hybridizes with A+C, A+B and A+B+C fragments in addition to the A fragment in the blot of the partially digested DNA (Figure 3, right). From this hybridization pattern, it can be easily concluded that these three fragments are arranged in the following order: B–A–C. Another method is Southern blotting with cloned markers (Figure 4). This procedure is useful for mapping restriction sites of another rare-cutting enzymes. Using these methods, we can construct a functional restriction map of the chromosome, and localize many genes on the chromosome.

3. GENERAL PROCEDURE

3.1 Strategy for Mapping

A linking library consists of genomic DNA fragments which contain a specific rare-cutting restriction enzyme site. Such clones are very useful as probes in PFGE and in mapping and cloning of large regions of DNA. Among rare-cutting restriction enzymes, *Not*I is convenient for the cloning, since there are many commercially available vectors containing a *Not*I cloning site. Other rare-cutting restriction enzymes such as *Asc*I and *Sse*8387I are also useful. Plasmid vectors of pNEB193 and pUC series contain an *Asc*I restriction site and an *Sse*8387I restriction site, respectively, in their multi-cloning site (Table 1).

We describe two methods using plasmid vectors and λ phage vectors in this chapter. The procedure using plasmid vectors is simple and is suitable for constructing linking libraries of bacterial genomes (2–5 Mb in size) [6,7]. On the other hand, the procedure using λ phage may be suitable for constructing linking libraries of mammalian genomes [8,11].

3.2 Production of Large DNA Fragments by Rare-cutting Restriction Enzymes

Many 8 bp-cutting restriction enzymes are now commercially available (Table 1). Among these enzymes, *Not*I and *Sfi*I have frequently been used for generating large fragments of mammalian DNA [19].

Bacterial genomes vary widely in their base composition. The G+C contents of *Staphylococcus*, *E. coli* and *Pseudomonas* are 34%, 50% and 57%, respectively [22]. Therefore, the number and size of fragments produced by digestion of these genomes with a restriction enzyme differ with different genomes. For digestion of bacterial genomes, various restriction enzymes should be tested.

The average size of restriction fragments generated by digestion of mammalian DNA with a restriction enzyme cannot simply be calculated on the basis of the size of the recognition sequence. Base composition and methylation pattern of the DNA sample as well as the methylation sensitivity of the restriction enzyme will influence the average size of the fragments. The well-known methylation site in mammalian cells is the sequence CpG. It is estimated that over 50% of the CpGs are methylated [23]. Some restriction enzymes including *Not*I are inhibited by the presence of 5-methylcytosine. Digestion of mammalian DNA samples with these enzymes will result in imcomplete cutting. This problem is discussed in Section 5.2.

Genomic DNAs are usually prepared in an agarose block. It is known that agarose is contaminated with small amounts of impurities that inhibit restriction enzymes. However, agarose of special grade, such as InCert agarose (FMC), prepared for cutting with restriction enzymes is available from many manufacturers.

3.3   Separation of Large DNA Fragments by PFGE and Southern Blot Analysis with Linking Clones

The principle of PFGE is discussed in Chapter 11. There are several parameters that must be considered before performing an electrophoretic separation of very high-molecular weight DNA. The separation of large DNA molecules by PFGE in an agarose gel is affected by agarose concentration, buffer concentration, temperature, switch times, voltage and total electrophoresis run time. In particular, the size of molecules that can be resolved is dependent largely upon the pulse time. In PFGE, DNA molecules are subjected to alternating the electric field imposed for a period of time called the switch time. When the field is switched, the DNA molecules are able to change direction or reorient in the gel matrix. Larger molecules take longer to reorient and have less time to move during each pulse, so that they migrate slower than smaller molecules. Resolution will be optimal for DNA molecules with reorientation time comparable to the pulse time. As the DNA size increases, the pulse time needed for resolving the molecules increases [18].

Large DNA fragments separated by PFGE are transferred onto membranes and detected by Southern hybridization analysis. The procedures that are described for Southern transfer and hybridization of DNA from standard agarose gels onto nylon or nitrocellulose membranes [20] are applicable to the analysis of large DNA fragments separated by PFGE.

When linking clones are used as probes in hybridization of large DNA fragments separated by PFGE, each clone will hybridize with two DNA fragments adjacent on the genome (Figure 2) [19].

3.4 Mapping with Cloned Markers

Once a physical map is constructed, various genes can be located on the map by simple blotting and hybridization. Furthermore, by partial mapping methods such as the Smith-Birnstiel procedure [24], the location of various restriction enzyme sites can be easily placed on the map.

3.5 Preparation of Solutions

TE buffer
    10 mM Tris-HCl, pH 8.0
    1 mM EDTA
Ligation buffer
    50 mM Tris-HCl, pH 7.5
    10 mM $MgCl_2$
    10 mM DTT
    50 µg mL^{-1} bovine serum albumin (BSA)
*Not*I buffer
    150 mM NaCl
    6 mM Tris-HCl, pH 7.5
    6 mM $MgCl_2$
    6 mM 2-mercaptoethanol (2-ME)
    0.01% Triton X-100
    0.1 mg/mL BSA
L-broth
    10 g Bacto-tryptone (Difco)

5 g Yeast extract (Difco)

5 g NaCl

Make up 1 L with distilled water, adjust to pH 7 and autoclave.

X-gal agar plate

L-broth containing 15 g L^{-1} Bacto agar (Difco)

50 μg mL^{-1} ampicillin

0.2 mM 5-bromo-4-chloro-3-indolyl-β-D-galactopyranoside (X-gal)

0.1 mM isopropyl-β-D-thiogalactopyranoside (IPTG)

Elution buffer

0.5 M ammonium acetate, pH 8.0

10 mM magnesium acetate

1 mM EDTA

0.1% SDS

Pett VI buffer

1 M NaCl

10 mM Tris-HCl, pH 7.5

EC lysing solution

1 M NaCl

0.1 M EDTA

6 mM Tris-HCl, pH 7.5

0.5% sodium lauroylsarcosine

0.2% deoxycholate

1 mg mL^{-1} lysozyme

20 μg mL^{-1} RNase

ESP solution

0.5 M EDTA, pH 9.0

1% sodium lauroylsarcosine

1 mg mL^{-1} proteinase K

Phosphate buffered saline (PBS)

8 g NaCl

0.2 g KCl

1.44 g Na$_2$HPO$_4$

0.24 g KH$_2$PO$_4$

Dissolve in about 800 mL of distilled water. Adjust to pH 7.4 with HCl and add distilled water to 1 L

PMSF

1 mM phenyl methyl sulphonyl fluoride (PMSF) in TE buffer

20 × SSC
  3 M NaCl
  0.3 M sodium citrate, pH 7.0
PFGE buffer
  0.1 M Tris
  0.1 M boric acid
  0.02 M EDTA

The solutions used for Southern hybridization are listed in Chapter 1 of Volume 1.

3.6   Additional Items

PFGE apparatus,
  e.g. Pulsaphor (Pharmacia)
        CHEF-DRII (Bio Rad)
        Hex-A-Field (BRL)
Absorbent paper, e.g. Whatman 3MM
Nylon membrane, e.g. Hybond N (Amersham)
Disposable column, e.g. Sephadex G-50 (Pharmacia)
Razor blade
Dialysis tubing

4.   METHOD

4.1   *Not*I Linking Clones: Using a Plasmid Vector

The following procedure is modified from that described by Okahashi *et al.* [6] and Okada *et al.* [7]. The scheme of this procedure is described in Figure 5.

1.   Digest chromosomal DNA (10–50 µg) with a 2- to 3-fold excess of a suitable 6 bp restriction enzyme such as *Eco*RI. Remove an aliquot and analyse the extent of digestion by agarose gel electrophoresis.

2.   When digestion is completed, extract the sample with phenol:chloroform, and precipitate the DNA with 2.5 volumes of ethanol for 30 min at −20°C. Recover the DNA by centrifugation at 10 000 g for 10 min at 4°C in a microfuge, and

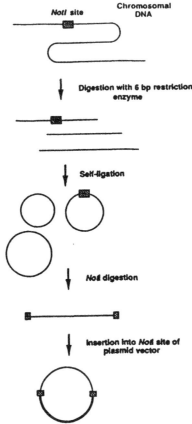

Figure 5. Construction of a *Not*I linking library with a plasmid vector. The chromosome is digested with a 6 bp cutting restriction enzyme, and then self-ligated. The ligated products are then digested with *Not*I and ligated with the *Not*I-digested plasmid vector. The sample is transformed into an -complementable *lac*Z mutant of *E. coli.* Lac⁻ transformants are identified on X-gal plates.

   redissolve it in ligation buffer. Adjust the DNA concentration to 1–5 µg mL⁻¹.

3. Add T4 DNA ligase (0.5 Weiss unit per 10 µL), and incubate overnight at 16°C.

4. Heat at 68°C for 10   min to inactivate the T4 DNA ligase. Precipitate the DNA with ethanol, and dissolve it in *Not*I buffer.

5. Add a 2- to 3-fold excess of *Not*I and digest the DNA for 2 h at 37°C. The fragments which contain *Not*I restriction sites are

linearized by this step.

6. Extract the sample with phenol:chloroform, precipitate the DNA with ethanol and then dissolve it in TE buffer.

7. Digest the plasmid vector pBluescript II (5 µg) with *Not*I (10 units) for 2 h at 37°C and then treat with calf intestinal alkaline phosphatase (1 unit) for 30 min at 37°C [20].

8. Set up the following ligation. Add 500 ng of the self-circularized, *Not*I-digested chromosomal DNA (step 6) plus 100 ng of the dephosphorylated *Not*I-digested pBluescript II (step 7) to 20 µL of the ligation buffer. Add 0.5 Weiss unit of T4 DNA ligase, and incubate overnight at 16°C.

9. Transform competent *E. coli* XL1-blue cells with the ligated plasmid from step 8 and plate on agar containing ampicillin (50 µg mL-¹), X-gal (0.2 mM) and IPTG (0.1 mM).

10. Isolate plasmids from the Lac⁻ transformants and determine their size and restriction map. These plasmid DNAs are used as *Not*I linking probes for hybridization to the *Not*I restriction fragments of the genome.

4.2   *Not*I Linking Clones: Using a Phage Vector

In mammalian DNAs, a chromosome-specific λ phage vector library is very useful for isolation of *Not*I linking clones. The following is a modification of the procedure of Saito et al. [11]. In their report, a *Not*I linking library was constructed by *supF* rescue from a Charon 21A library containing *Hin*dIII inserts prepared from flow-sorted human chromosome 21 [8,11]. The *supF* gene was isolated from πvχ plasmid [11]. The scheme of this procedure is described in Figure 6.

1. Digest a plasmid containing the *supF* gene with *Not*I. Separate the DNA fragments by polyacrylamide gel electrophoresis.

2. Using a sharp razor blade, cut out the gel segment containing the band of *supF* (210 bp).

3. Crush the gel slice, add 2 volumes of elution buffer, and extract at 37°C for 3–4 h.

4. Centrifuge the sample at 15 000 g for 5 min at 4°C in a microfuge.

5. Collect the supernatants, and remove any residual pieces of

polyacrylamide gel by passing the supernatant through a disposable column.

6. Recover the DNA by ethanol precipitation, and redissolve it in TE buffer.

7. Prepare a λ library of the target chromosome. Some chromosomal libraries of mammalian cells can be obtained from the American Type Culture Collection (ATCC). Amplify this λ library by a plate lysate method, and prepare purified λDNA [20; see Chapter 8].

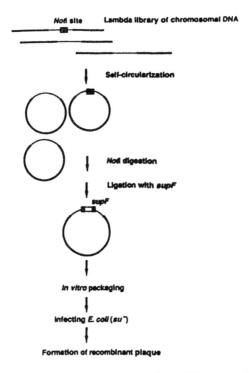

Figure 6.   Construction of a *NotI* linking library with a λ phage vector. Linking fragments are isolated by preparing a total genomic library or a library from sorted individual chromosome in a λ phage vector. DNA is isolated from an amplification of this library. The DNA is self-circularized, and digested with *NotI*. The suppressor tRNA (*supF*) with *NotI* linkers is inserted. The DNA is packaged by using a packaging kit. *E. coli* (Su⁻) is infected with phages. Phages which contain the *supF* gene form plaques.

8. Dilute the λ DNA samples with ligation buffer base (50 mM Tris-HCl, 10 mM MgCl$_2$, pH 7.5) to a final concentration of 1–2 µg mL^{-1}. The total volume of the reaction mixture should be 0.5 mL.

9. Heat the samples at 68°C for 10 min.

10. Add ATP, DTT and T4 DNA ligase to final concentrations of 1 mM, 10 mM and 0.5 Weiss unit per 10 µL, respectively, and ligate overnight at 15°C.

11. Inactivate the T4 DNA ligase by heating the sample at 68°C for 10 min.

12. Add 30 units of *Not*I and 90 µL of *Not*I buffer, and incubate overnight at 37°C.

13. Extract the sample with phenol:chloroform.

14. Dialyse the sample against TE buffer.

15. Prepare the ligation mixture containing λ DNA digested with *Not*I (a final concentration of 2 µg mL^{-1}) and *supF* fragment (a final concentration of 0.01 µg mL^{-1}).

16. Incubate the reaction mixture overnight at 15°C.

17. Extract the DNA sample with phenol:chloroform, precipitate the DNA with ethanol, and redissolve it in TE buffer.

18. Package the sample with an *in vitro* packaging kit. *In vitro* packaging kits are commercially available (e.g. Giga Pack, Stratagene).

19. Infect *E. coli* CA274 (HfrC *lac*$_{125am}$*trp*$_{am}$) with the packaging lysate.

20. Plate the infected *E. coli* on an X-gal plate, and incubate at 37°C to allow plaque formation. Select blue plaques.

21. Isolate the λ DNA from each recombinant clone (see Chapter 8).

4.3 Preparation of Chromosomal DNA: Bacteria

Procedures for the preparation of chromosomal DNA from gram-negative and gram-positive bacteria contain a step to lyse bacteria by various lytic enzymes (cell wall degrading enzymes). Several bacteria are resistant to any lytic enzymes. Furthermore, many bacteria produce capsular-like polysaccharides. These substances sometimes remain in the agarose block and inhibit restriction enzymes. Therefore, it is important to select bacterial species and to establish the condition for cell lysis.

In this section, a procedure for preparing chromosomal DNA from *E. coli* is described [2,7,18].

1. Grow 10 mL of *E. coli* culture to $OD_{550} = 0.5$. Centrifuge the culture at 6000 g for 10 min, and collect the cells.
2. Suspend the cells in 10 mL of PettIV buffer.
3. Mix the cell suspension with an equal volume of 1.4% low-melting-temperature agarose (prepared in sterile water and cooled to 45°C). Transfer immediately with a pipette to a mold to make a 100 μL block.
4. Allow the samples to harden on ice for 15 min.
5. Remove the sample blocks from the mold, and incubate in EC lysing solution overnight at 37°C with constant gentle shaking.
6. Incubate the blocks in ESP solution for 2 days at 50°C with gentle shaking.
7. The samples may then be stored at 4°C in the ESP solution.

4.4 Preparation of Chromosomal DNA: Mammalian Cells

Preparation of DNA from mammalian cells in agarose blocks is easier than solution procedures [18,19]. In chromosome mapping of mammalian cells, a hybrid cell line is very valuable. For example, a human-mouse hybrid cell line with chromosome 21 is used as a source of genomic DNA for construction of a chromosome 21 restriction map [8]. The restriction map can be constructed by using Alu banding.

1. Collect mammalian cells, and resuspend in PBS to a final concentration of 1–2×10^7 cells per mL.
2. Mix the cell suspension with an equal volume of 1.4% low-melting-temperature agarose (prepared in PBS and cooled to 45°C). Transfer immediately with a pipette into a mold.
3. Allow the samples to harden on ice for 15 min.
4. Remove the sample blocks from the mold, and incubate in ESP solution for 2 d at 50°C with gentle shaking.
5. The samples may then be stored at 4°C in the ESP solution.

4.5 *NotI* Digestion of Chromosomal DNA in Agarose

1. Incubate the agarose blocks containing chromosomal DNA at room temperature in 50 volumes of TE buffer containing 1 mM PMSF for 2 h. Repeat this step.

2. Wash the agarose blocks in 50 volumes of TE buffer three times for 2 h at room temperature.
3. Transfer the sample blocks to microfuge tubes, and add 10 volumes of the *Not*I buffer. Incubate the samples for 30 min at 4°C.
4. Remove the buffer and add 4 volumes of fresh *Not*I buffer. Add *Not*I (5 units per 1 µg DNA), and incubate overnight at 37°C.
5. Wash the samples in 5 volumes of ESP buffer for 1 h at 50°C.
6. The samples may be stored at 4°C until used for PFGE analysis.

4.6 PFGE and Southern Transfer

As described in Section 3.3, resolution of DNA fragments by PFGE is dependent on pulse time, electrophoresis run-time and temperature [19]. Conditions of PFGE can be chosen to optimize the resolution of a particular size range of molecules (see Chapters 11 and 13 for additional information on PFGE conditions).

Since DNA fragments larger than 20 kb cannot be transferred efficiently by Southern blotting, DNA fragments separated by PFGE must be cleaved before transfer onto a membrane. The DNA can be cleaved either by using acid (0.25 M HCl) or by UV irradiation.

After PFGE, the gel is stained with 1 µg mL^{-1} of ethidium bromide for 10 min with constant agitation, and photographed on a UV illuminator. To break large DNA molecules into smaller fragments, the gel is incubated in 0.25 M HCl for 10 min at room temperature. The gel is then denatured in 0.5 M NaOH containing 1.5 M NaCl for 1 h, and neutralized with 0.5 M Tris containing 1.5 M NaCl (pH 7.5) for 1 h with gentle agitation. The gel is blotted to a nylon membrane by ascending transfer overnight with 20 × SSC. The membrane is then fixed by UV irradiation as indicated by the manufacturer, and used for hybridization [20].

5. GENERAL POINTS

5.1 Characterization of Linking Clones

As shown in Figures 5 and 6, the procedure for constructing linking clones contains a self-circularization step. In the self-circularization step, ligation between exogenous fragments always occurs. Therefore, characterization of the linking clones obtained is needed, since parts of the recombinant clones obtained may contain DNA fragments that are not adjacent in the genome.

It is difficult to construct a complete set of linking clones. Some clones may be deleted during expansion of the clones because of a toxic effect of the cloned sequence in *E. coli*. Linking clones should hybridize specifically with only two adjacent *Not*I fragments. However, some linking clones hybridize with three or more *Not*I fragments. This may come from the existence of repeating sequence units in the genome. For example, the hybridization by the *Not*I-linking clones in *Shigella flexneri* is due to the existence of insertion elements such as IS*1* or IS*4* scattered on the chromosome [7].

5.2   Effect of Methylation of DNA

The recognition sequence of *Not*I (GCGGCCGC) contains two CpG sequences potentially accessible to mammalian methylases. Methylated sequences are not cleaved with *Not*I. The effect of methylation should be considered when a physical map of mammalian cells is constructed.

There are both tissue-specific and cell-specific differences in the level of methylation. Alterations in the level of methylation may occur with the age of the organism, hormone treatment, or malignancy. Thus, it is difficult to clear up the cause for incomplete digestion. Hybridization patterns must be cautiously interpreted when different DNA sources are used, because apparent polymorphisms may merely reflect variations in the state of methylation. Differences in methylation among cell lines are reported by Ichikawa et al. [10].

Physical mapping is a very rapid developing area in molecular biology. Many recent articles have described physical maps of mammalian genomes. In these articles, cloning with YAC vectors is the most common procedure for the construction of a clone map. However, the procedure for construction of a *Not*I map with linking clones is also utilized for constructing the YAC clone maps. This method can be applied to the detailed mapping of cloned YAC fragments. In addition to the mapping of mammalian genome, physical maps of many bacterial genomes can be easily constructed by using linking clones.

6. TROUBLE SHOOTING

6.1 Low Molecular Weight Smears in Southern Hybridization

The DNA sample is degraded by nuclease contamination during the sample preparation and/or during the restriction digestion. Prepare a new DNA sample for PFGE.

6.2 Broad and Unclear Bands Detected in Southern Hybridization

Incorrect electrophoresis conditions were used. Check the temperature and electrophoresis condition such as the voltage and the pulse time, or remake the electrophoresis buffer.

6.3 Detection of Three or More Hybridization Bands

Check the linking clone used as the probe. Or the probe may contain a repetitive sequence in the genome (see Section 5.1).

REFERENCES

[1] Kohara, Y., Akiyama, K and Isono, K. The physical map of the whole *E. coli* chromosome: Application of a new strategy for rapid analysis and sortin of a large genomic library. Cell, 50 (1987) 495–508.

[2] Smith, C. L., Econome, J. G., Schutt, A., Klco, S. and Cantor, C. R. A. Physical map of the *Escherichia coli* K12 genome. Science, 236 (1987) 1448–1453.

[3] Carle, G. F. and Olson, M. V. Separation of chromosomal DNA molecules from yeast by orthogonal-field-alternation gel electrophoresis. Nucleic Acids Res., 12 (1984) 5647–5664.

[4] Oliver, S. G., van der Aart, J. M., Agostoni-Carbone, M. L., *et al.* The complete DNA sequence of yeast chromosme III. Nature, 357 (1992) 38–46.

[5] Sulston, J., Du, Z., Thomas, K., Wilson, R., Hiller, L., Staden, R., *et al.* The *C. elegans* genome sequencing project. Nature, 356 (1992) 37–42.

[6] Okahashi, N., Sasakawa, C., Okada, N., Yamada, M., Yoshikawa, M., Tokuda, M., Takahashi, I. and Koga, T. Construction of a *Not*I restriction map of the *Streptococcus mutans* genome. J. Gen. Microbiol., 136 (1990) 2217–2223.

[7] Okada, N., Sasakawa, C., Tobe, T., Talukder, K. A., Komatsu, K. and Yoshikawa, M. Construction of a physical map of the chromosome of *Shigella fleneri* 2a and the direct assignment of nine virulence-associated loci identified by Tn*5* insertions. Mol. Microbiol., 5 (1991) 2171–2180.

[8]   Ichikawa, H., Shimizu, K., Saito, A., Wang, D., Oliva, R., Kobayashi, H. *et al.*
      Long-distance restriction mapping of the proxiaml long arm of human
      chromosome 21 with *Not*I linking clones. Proc. Natl. Acad. Sci. USA, 89
      (1992) 23–27.

[9]   Gardiner, K., Horisberger, M., Kraus, J., Tantravahi, U., Korenberg, J., Rao,
      V., Reddy, S. and Patterson, D. Analysis of human chromosome 21:
      Correlation of physical and cytogenetic maps; gene and CpG island
      distributions. EMBO J., 9 (1990) 25–34.

[10] Pohl, T. M., Zimmer, M., MacDonald, M. E., Smith, B., Bucan, M., Pouska, A.
      et al. Construction of a *Not*I linking library and isolation of new markers close
      to the Huntigton's disease gene. Nucleic Acids Res., 16 (1988) 9185–9198.

[11] Saito, A., Abad, J. P., Wang, D., Ohki, M., Cantor, C. R. and Smith, C. L.
      Construction and characterization of a *Not*I linking library of human
      chromosome 21. Genomics, 10 (1991) 618–630.

[12] Burke, D. T., Carle, G. F. and Olson, M. V. Cloning of large segments of
      exogenous DNA into yeast by means of artificial chromosome vectors.
      Science, 236 (1987) 806–812.

[13] Brownstein, B. H., Silverman, G. A., Little, R. D., Burke, D. T., Korsmeyer, J.,
      Schlessinger, D. and Olson, M. V. Isolation of single-copy human genes from
      a library of yeast artificial chromosome clones. Science, 244 (1989)
      1348–1351.

[14] Green, E. D. and Olson, M. V. Chromosomal region of the cystic fibrosis gene
      in yeast artificial chromosomes: A model for human genome mapping.
      Science, 250 (1990) 94–98.

[15] Foote, S., Vollrath, D., Hilton, A. and Page, D. C. The human Y chromosome:
      Overlapping DNA clones spanning the euchromatic region. Science, 258
      (1992) 60–66.

[16] Chumakov, I., Rigault, P., Guillou, S., et al. Continuum of overlapping clones
      spanning the entire human chromosome 21q. Nature, 359 (1992) 380–387.

[17] Schwartz, D. C. and Cantor, C. R. Separation of yeast chromosome-sized
      DNAs by pulsed field gradient gel electrophoresis. Cell, 37 (1984) 67–75.

[18] Smith, C. L. and Cantor, C. R. Purification, specific fragmentation, and
      separation of large DNA molecules. Methods Enzymol., 155 (1987) 449–466.

[19] Smith, C. L., Lawrance, S. K., Gillespie, G. A., Cantor, C. R., Weissman, S.
      M. and Collins, F. S. Strategies for mapping and cloning macroregions of
      mammalian genomes. Methods Enzymol. 151 (1987) 461–489.

[20] Sambrook, J., Fritsch, E. F. and Maniatis, T. *Molecular Cloning. A Laboratory
      Manual.* Cold Spring Harbor Laboratory Press, Cold Spring Harbor, N.Y.,
      1989.

[21] Wallace, M. R., Fountain, J. W., Brereton, A. M. and Collins, F. S. Direct
      construction of a chromosome-specific *Not*I linking library from flow-sorted
      chromosomes. Nucleic Acids Res., 17 (1989) 1665–1677.

[22] McClelland, M., Jones, R., Patel, Y. and Nelson, M. Restriction endonicleases
      for pulsed field mapping of bacterial genomes. Nucleic Acids Res., 15 (1987)
      5985–6005.

[23] Bird, A., Taggart, M., Frommer, M., Miller, O. J. and Macleod, D. A fraction
      of the mouse genome that is derived from islands of nonmethylated, CpG-rich
      DNA. Cell, 40 (1985) 91–99.

[24] Smith, H. O. and Birnstiel, M. L. A simple method for DNA restriction site
      mapping. Nucleic Acid Res., 3 (1976) 2387–2398.

Chapter 13

PULSED-FIELD GEL ELECTROPHORESIS
ISOLATION AND ANALYSIS OF LARGE LINEAR PLASMIDS

Haruyasu Kinashi

OUTLINE

Methods in Gene Technology, Volume 2, pages 227–239
Copyright © 1994 JAI Press Ltd
All rights of reproduction in any form reserved.
ISBN: 1-55938-264-3

1. INTRODUCTION

There are several cases where the genetic evidence suggests that a character is encoded by a plasmid but no plasmid can be physically detected. Some of these phenomena could be explained by involvement of a large linear plasmid. So far, large linear plasmids have not been separated by conventional techniques from large linear fragments of chromosomal DNA formed during DNA preparation. Pulsed field gel electrophoresis (PFGE) [1–3; see also Chapters 11, 12] has enabled for the first time separation of large linear plasmids from chromosomal DNA fragments.

The following are examples of large linear plasmids which have been confirmed to have a special function in microorganisms. In *Nocardia opaca*, a soluble dehydrogenase responsible for the hydrogen autotrophy is encoded by a large linear plasmid [4]. A 49-kb linear plasmid was reported to contain the *ospA* and *ospB* genes which encode the major outer membrane proteins of a spirochete, *Borrelia burgdorferi* [5]. The biosynthetic gene cluster for an antibiotic methylenomycin is carried on a 350-kb giant linear plasmid, SCP1, in *Streptomyces coelicolor* [6]. SCP1 also carries the *sapC*, *sapD*, and *sapE* genes which encode the spore-associated proteins of this strain [7]. Although the number of examples is not many, it will increase rapidly by the use of PFGE.

Following the isolation of SCP1 from *S. coelicolor*, we detected giant linear plasmids from several antibiotic-producing strains of *Streptomyces* species [8]. *Streptomyces* are mycelial, spore-forming, Gram-positive bacteria which produce a variety of antibiotics and bioactive compounds. This genus also characteristically exhibits a very high degree of genetic instability [9,10]. This property might be explained by the chromosomal DNA being readily rearranged, or by involvement of a large linear plasmid. The mechanisms of genetic instability of *Streptomyces* are now being studied extensively by several groups using PFGE analysis. We have proposed a model for integration of SCP1 into the *Streptomyces coelicolor* chromosome by examining the junctions of integration [11,12]. This chapter describes methods for the isolation and analysis of large linear plasmids from *Streptomyces*. These methods can also be applied to the isolation and analysis of chromosomal DNA.

2. PREPARATION OF TOTAL DNA FOR PFGE

Large DNAs are very sensitive to physical shearing during the usual methods for DNA preparation in solution. Therefore, in order to obtain intact chromosomes from *Saccharomyces cerevisiae*, Schwartz and Cantor [1] embedded yeast cells in low melting temperature agarose and made protoplasts by treating with Zymolyase *in situ*. The protoplasts were then lysed by adding a solution containing proteinase K and sodium N-lauroyl sarcosine. However, direct application of this method seemed unlikely to be suitable for *Streptomyces*, because *Streptomyces* are bulky filamentous bacteria and their protoplast formation usually gives a lot of cell wall debris, which might interfere with both separation of DNA on the PFGE gel and digestion of DNA by restriction endonucleases. Therefore, we made protoplasts in solution, which were then passed through a glass wool filter, collected by centrifugation, embedded in agarose gel and treated with pronase and SDS [16].

After we published this protoplast method, several groups reported a mycelium method where *Streptomyces* mycelium was directly embedded in agarose and used successfully to get chromosomal DNA [15,17,18]. I describe here both methods with some modifications from the original ones and discuss both their merits and demerits.

2.1 Protoplast Method

2.1.1 Buffers and Reagents

Yeast Extract-Malt Extract Medium (YEME) [13]

Difco Yeast Extract	3 g
Difco Bacto-Peptone	5 g
Oxoid Malt Extract	3 g
Glucose	10 g
Sucrose	340 g

Make up to one litre and autoclave

After autoclaving add:

$MgCl_2.6H_2O$ (2.5 M)   2 mL/L

For preparing protoplasts, also add:

Glycine (20%)   25 mL/L

P3 Buffer [14]

70 mM NaCl

5 mM $MgCl_2$

5 mM $CaCl_2$

10.3% Sucrose

5.73% TES: N-tris(hydroxymethyl)methyl-2-aminoethanesulphonic acid, pH 7.2

PWP Buffer [14]

70 mM NaCl

10 mM $MgCl_2$

20 mM $CaCl_2$

10.3% Sucrose

5.73% TES, pH 7.2

P (Protoplast) Buffer [13]

Make up the following base solution:

Sucrose	103 g
K_2SO_4	0.25 g
$MgCl_2.6H_2O$	2.02 g
Trace Element Solution	2 mL
Distilled Water to 800 mL	

Dispense in 80 mL aliquots and autoclave

Before use, add to each flask in order:

KH_2PO_4 (0.5%)	1 mL
$CaCl_2.2H_2O$ (3.68%)	10 mL
TES Buffer (5.73% adjusted to pH 7.2)	10 mL

Trace Element Solution

Per litre:

$ZnCl_2$	40 mg
$FeCl_3.6H_2O$	200 mg
$CuCl_2.2H_2O$	10 mg
$MnCl_2.4H_2O$	10 mg
$Na_2B_4O_7.10H_2O$	10 mg
$(NH_4)_6Mo_7O_{24}.4H_2O$	10 mg

TE25Suc [15]
   25 mM Tris-HCl, pH 8.0
   25mM EDTA, pH 8.0
   10.3% Sucrose

Proteinase K: Merck, Rahway, N.J., 1 mg/mL in TE25Suc.

Low melting point (LMP) agarose: Agarose type VII, Sigma, St. Louis, Mo.

0.5 M EDTA, pH 8.0

Pronase: Actinase E, Kaken Pharmaceuticals, Tokyo.
   Dissolve the enzyme (5 mg/mL) in TE25Suc and incubate at 37°C for 2 h to digest contaminated nucleases.

10% sodium dodecyl sulphate (SDS)

2.1.2 Procedure

1. Grow a *Streptomyces* strain in YEME medium[a] supplemented with 0.5% glycine[b] in a baffled Erlenmeyer flask on a rotary shaker at 30°C to early stationary phase.
2. Wash the mycelium twice in 10.3% sucrose by centrifugation at 4000 r.p.m. for 10 min. The washed mycelium can be stored in a freezer until use.
3. Suspend about 5 mL of the wet mycelium in 25 mL of P3 buffer[c], add 60 mg of lysozyme (final 2 mg/mL), and incubate at 30°C for 10–30 min with gentle shaking. Check the protoplast formation by periodic microscopic observations.
4. Filter the protoplasts through a glass wool filter, wash out the filter with PWP buffer and collect the protoplasts by centrifugation at 3500 r.p.m. for 10 min.
5. Resuspend the protoplast pellet gently in 1.5 mL of PWP buffer and transfer to a Falcon 3002 plate (diameter of 6 cm). The volume of PWP buffer should be reduced when the protoplast pellet is small.
6. To the protoplast suspension, add 3 mL of molten 1.5% low melting point (LMP) agarose (final 1.0%) in PWP kept at 40°C and mix gently.
7. Solidify the agarose for 10 min in a refrigerator.
8. Overlay with 1 mL of 0.5 M EDTA and leave the plate at 4°C for 1 h.
9. Add 4 mL of a pronase solution[d] (5 mg/mL) in TE25Suc and 1 mL of 10% SDS solution and incubate at 50°C overnight.

10. Replace the buffer by 0.5 M EDTA and store the plate at 4°C
     until use. Replace the buffer by 0.5 M EDTA several times
     during storage.

2.1.3 Notes

a. YEME medium is suitable for protoplast formation. However, any medium
    can be used if mycelium grown in it is susceptible to lysozyme.
b. Addition of glycine makes the protoplast formation easy but also inhibits the
    growth. The appropriate concentration of glycine is around 0.5% but is
    different from strain to strain.
c. We usually use P3 and PWP buffers [14] for protoplasting. P3 buffer is
    suitable for protoplast formation of lysozyme-resistant *Streptomyces* species,
    while the protoplasts formed are stable in PWP buffer. P buffer [13] can also
    be used in place of P3 and PWP buffers.
d. Use a proteinase K solution (1 mg/mL) in TE25Suc in place of the pronase
    solution if necessary. Pronase is a mixture of several proteases from
    Streptomyces griseus, while proteinase K is serine-protease from *Tritirachium
    album* and its activity is inhibited by phenymethylsulphonyl fluoride (see
    Chapter 1). Therefore, it is better to use proteinase K when total DNA in
    agarose gel is directly digested with restriction enzymes in the next step. Since
    proteinase K is expensive, it can be replaced by pronase in appropriate cases.
    For example, pronase can be used for isolation of large linear plasmids as
    described here, because proteins do not move on the PFGE gel and do not
    contaminate the plasmid band. However, Leblond *et al.* [17] successfully used
    pronase for digestion analysis of the chromosomal DNA of *Streptomyces
    ambofaciens*.

2.2 Mycelium Method

2.2.1 Buffers and Reagents

TE25Suc [15]
  25 mM Tris-HCl, pH 8.0
  25 mM EDTA, pH 8.0
  10.3% Sucrose

Low melting point (LMP) agarose: Agarose type VII, Sigma,
St. Louis, Mo.

Lysozyme
  2 mg/mL in TE25Suc

Pronase: Actinase E, Kaken Pharmaceuticals, Tokyo. Dissolve the
enzyme (5 mg/mL) in TE25Suc and incubate at 37°C for 2 h to digest
contaminated nucleases.

10% sodium dodecyl sulphate (SDS)

0.5 M EDTA, pH 8.0

2.2.2   Procedure

1. Mix 1 mL of the wet mycelium and 0.5 mL of TE25Suc in a Falcon 3002 plate. When the total DNA sample is directly digested with restriction endonucleases, a much smaller amount of mycelium should be used for complete digestion.
2. To the mycelium suspension, add 3 mL of molten 1.5% LMP agarose in TE25Suc and mix well until the mycelium is not precipitated on the bottom.
3. Solidify the agarose at 4°C for 10 min.
4. Add 4 mL of lysozyme (2 mg/mL) in TE25Suc and incubate at 37°C for 3–5 h. The agarose gel should be detached from the bottom of the plate and floated in the buffer during protoplasting.
5. Replace the lysozyme solution by 1 mL of 0.5 M EDTA, 4 mL of pronase (5 mg/mL) in TE25Suc and 1 mL of 10% SDS, and incubate at 50°C overnight.
6. Replace the buffer by 0.5 M EDTA and store the plate at 4°C until use.

2.3   Considerations

The two methods described above have the following properties. The protoplast method is able to obtain a more concentrated DNA than the mycelium method. This property is necessary for the physical mapping of a linear plasmid, because in some cases restriction fragments of a linear plasmid isolated by the second PFGE and further subjected to the second digestion and then to the third PFGE. However, this method is more tedious and sometimes causes DNA degadation which give a smear band ranging from 50 to 200 kb on the PFGE gel. This smear band makes difficult the detection of a linear plasmid or digested chromasomal fragments of the same size on the PFGE gel (degraded DNAs can be removed as described later).

On the other hand, the mycelium method usually gives more intact DNA. However, this method is not suitable for obtaining a quite high DNA concentration due to the bulkiness of mycelium. This method does not work well when the mycelium forms pellets during shaking culture. The pellets should be disrupted by homogenization before use. Although both methods are working well in the author's laboratory, the mycelium method is recommended as the method of first choice for its ease of handling. When this method does not work well or there is need for a quite concentrated DNA for the physical mapping of a plasmid and so on, try the protoplast method

3.   ISOLATION OF LARGE LINEAR PLASMIDS BY PFGE

3.1   Background

Since the invention of PFGE by Schwartz and Cantor [1], several modifications of the PFGE system have been reported. Among them, contour-clamped homogeneous electric fields (CHEF) developed by Chu *et al.* [19] is now the most popular. The CHEF apparatus can be bought from several companies; for example, Pulsaphor from LKB, Germany and CHEF-DRIII from Bio-Rad, U.S.A. We now use a CHEF apparatus from Biocraft, Tokyo, Japan. It has a modification from the original machine in which the buffer is kept at 14°C by circulating cooling water under the gel box instead of circulating the buffer itself through a cooling unit. See Chapter 11 for a review of different forms of PFGE.

3.2   Buffers and Reagents

0.5 M EDTA, pH 8.0

0.5 × TBE Buffer
   44.5 mM Tris
   44.5 mM Boric acid
   1 mM EDTA

Low melting point (LMP) agarose
   1% in 0.5 × TBE

Ethidium bromide
   1 μg/mL in water

TE buffer
   10 mM Tris-HCl, pH 8.0
   1 mM EDTA, pH 8.0

3.3   Procedure

Isolation of large linear plasmid by PFGE can be done as follows:

1. Cut the total DNA gel sample (called insert) stored in 0.5 M EDTA into long strips of 5 mm width.
2. Rinse the gel strips for a while in 0.5 × TBE to remove EDTA.
3. Apply the gel strips to a long loading well on 1.0% LMP agarose gel in 0.5 × TBE.
4. Conduct the CHEF gel electrophoresis in 0.5 × TBE at 14°C at a voltage between 120 and 180 V for 1–2 d. The pulse time selected depends on the length of the plasmid DNA to be separated. The following empirical equation has been proposed for the relationship between the length of a linear DNA and a pulse time [20]:

   $\log_{10}L = 0.78\log_{10}E + 1.48$

   where L is the length of the linear DNA (kb), $E = TV/10 \times D$, E is the effective pulse time (s), T is the actual pulse time (s), V is the applied voltage and D is the length of the buffer chamber in the electrophoresis apparatus.
5. After gel electrophoresis, stain the gel in 1 μg/mL of ethidium bromide in water for 30 min and wash in water for 1 h to overnight with gentle shaking.
6. Detect the plasmid band under UV irradiation and excise it from the gel. Cut the plasmid band into small pieces (1–2 × 5 × 10 mm), wash them for 1 h twice in TE buffer, and store in TE buffer at 4°C.

It is most important to prevent nuclease contamination during isolation and analysis of DNA. For this purpose, wash all the apparatus and all plastic and glassware well by detergent and rinse thoroughly with distilled water; use gloves to handle them.

4. DIGESTION OF PLASMID AND CHROMOSOMAL DNA IN AGAROSE GEL

4.1 Buffers and Reagents

TE buffer
  10 mM Tris-HCl, pH 8.0
  1 mM EDTA, pH 8.0
0.5 M EDTA, pH 8.0
0.1 mM phenylmethylsulphonyl fluoride (PMSF)
  Make a 100 mM stock solution in 2-propanol and dilute with TE.

4.2   Procedure

A large linear plasmid isolated by PFGE can be digested in agarose gel as follows:

1. Wash the plasmid gel pieces stored in TE buffer in two changes of the appropriate reaction buffer, for 1 h each.
2. Digest the gel pieces overnight with a restriction endonuclease in the buffer (supplemented with 100 μg/mL of bovine serum albumin) with gentle shaking.
3. Stop the digestion by adding 0.5 M EDTA, replace the buffer by 0.5 M EDTA and store the gel pieces at 4°C until required for PFGE analysis.

When a total DNA sample is directly digested with restriction endonucleases, PMSF treatment should be done before digestion to inhibit proteinase K (or pronase) as follows:

1. Cut the insert gel into small pieces ($1-2 \times 5 \times 10$ mm) and wash them in TE for 1 h twice.
2. Treat the gel pieces with 0.1 mM PMSF for 1 h with gentle shaking.
3. Wash the gel pieces in TE for 1 h and in the reaction buffer for 1 h, successively.
4. Digest the gel pieces as described above.

Digested fragments of a large linear plasmid or chromosomal DNA thus obtained can be subjected to PFGE as described for isolation of large linear plasmids.

Note: If the insert contains a large amount of degraded DNA, pre-PFGE electrophoresis is effective. Namely, short degraded DNAs can be eliminated from the insert by CHEF with a short pulse time (e.g. 10 s) for a short period (e.g. 5 h). The insert recovered from the loading well contains almost only intact chromosomal DNA suitable for digestion.

5.   SDS-PFGE

Stam *et al.* [21] reported that intact protein-bound pGKL1 and pGKL2, linear plasmids from *Kluyveromyces lactis*, were able to move in SDS-agarose gel electrophoresis in contrast to normal agarose gel electrophoresis where protein-bound plasmids did not move. Most of the linear DNA genomes reported so far have a protein bound to the 5′-ends. Therefore, we tried CHEF in the presence of

SDS (named SDS-CHEF) to analyze a protein-binding to SCP1 and its restriction fragments [16].

SDS-CHEF was carried out by adding 0.2% SDS to both the buffer and agarose gel. We found that addition of SDS prolonged the pulse time needed to separate the same fragments; for example, we used pulse times of 24 s for CHEF and 48 s for SDS-CHEF, respectively to separate *Spe*I fragments of SCP1 [16].

Analysis of a protein binding to a linear plasmid is carried out as follows:

1. Prepare the pronase-untreated DNA insert by the same method described above except for the pronase treatment being omitted.
2. Subject the pronase-untreated insert to SDS-CHEF with LMP agarose; a protein-bound plasmid is able to move under these conditions.
3. Cut the plasmid band out of the gel and wash it in TE buffer to remove SDS.
4. Digest the pronase-untreated plasmid with restriction endonucleases as usual.
5. Analyze the pronase-treated and untreated plasmids and their restriction fragments by normal CHEF and SDS-CHEF and compare their mobilities.

   The untreated plasmid and its terminal fragment containing a protein should remain at the origin in normal CHEF, but move at the same speed with the pronase-treated counterpart in SDS-CHEF. The mode of protein binding to the ends of large linear plasmids can be analysed further by digestion with exonuclease III and lambda exonuclease [16, 22].

REFERENCES

[1] Schwartz, D.C. and Cantor, C.R. Separation of yeast chromosome-sized DNAs by pulsed field gradient gel electrophoresis. Cell, 37 (1984) 67–75.

[2] Smith, C.L. and Cantor, C.R. Purification, specific fragmentation, and separation of large DNA molecules. Methods Enzymol., 155 (1987) 449–467.

[3] Carle, G.F. and Olson, M.V. Orthogonal-field-alternation gel electrophoresis. Methods Enzymol., 155 (1987) 468–482.

[4] Kalkus, J., Reh, M. and Schlegel, G. Hydrogen autotrophy of *Nocardia opaca* strains is encoded by linear megaplasmids. J. Gen. Microbiol., 136 (1990) 1145–1151.

[5]   Barbour, A.G. and Garon, C.F. Linear plasmids of the bacterium *Borrelia burgdorferi* have covalently closed ends. Science, 237 (1987) 409–411.

[6]   Kinashi, H., Shimaji, M. and Sakai, A. Giant linear plasmids in *Streptomyces* which code for antibiotic biosynthesis genes. Nature, 328 (1987) 454–456.

[7]   McCormick, J.R., Santamaria, R. and Losick, R. The genes for three spore-associated proteins are encoded on linear plasmid SCP1 in *Streptomyces coelicolor*. Proceedings of the 8th International Symposium on Biology of Actinomycetes. Madison, Wisconsin, 1991, P 1–125.

[8]   Kinashi, H. and Shimaji, M. Detection of giant linear plasmids in antibiotic producing strains of *Streptomyces* by the OFAGE technique. J. Antibiot., 40 (1987) 913–916.

[9]   Chater, K.F., Henderson, D.J., Bibb, M.J. and Hopwood, D.A. Genome flux in *Streptomyces coelicolor* and other streptomycetes and its possible relevance to the evolution of mobile antibiotic resistance determinants. In: Kingsman, A.J., Kingsman, S.M. and Chater, K.F.(eds.), *Transposition* (Society for General Microbiology Symposium, 43), Cambridge University Press, Cambridge, 1988, pp.7–42.

[10] Huetter, R., Birch, A., Haeusler, A., Voegtli, M., Madon, J. and Krek, W. Genome fluidity in *Streptomyces*. In: Okami, Y., Beppu, T. and Ogawara, H.(eds.), *Biology of Actinomycetes '88*, Japan Scientific Societies Press, Tokyo, 1988, pp.111–116.

[11] Hanafusa, T. and Kinashi, H. The structure of an integrated copy of the giant linear plasmid SCP1 in the chromosome of *Streptomyces coelicolor* 2612. Mol. Gen. Genet., 231 (1992) 363–368.

[12] Kinashi, H., Shimaji-Murayama, M. and Hanafusa, T. Integaration of SCP1, a giant linear plasmid, into the *Streptomyces coelicolor* chromosome. Gene, 115 (1992) 35–41.

[13] Hopwood, D.A., Bibb, M.J., Chater, K.F., Kieser, T., Bruton, C.J., Kieser, H.M., Lydiate, D.J., Smith, C.M., Ward, J.M. and Schrempf, H., *Genetic Manipulation of Streptomyces: a Laboratory Manual*, The John Innes Foundation, Norwich, England, 1985.

[14] Shirahama, T., Furumai, T. and Okanishi, M. A modified regeneration method for streptomycete protoplasts. Agric. Biol. Chem., 45 (1981) 1271–1273.

[15] Kieser, M.H., Kieser, T. and Hopwood, D.A. A combined genetic and physical map of the *Streptomyces coelicolor* A3(2) chromosome. J. Bacteriol., 174 (1992) 5496–5507.

[16] Kinashi, H. and Shimaji-Murayama, M. Physical characterization of SCP1, a giant linear plasmid from *Streptomyces coelicolor*. J. Bacteriol., 173 (1991) 1523–1529.

[17] Leblond, P., Francou, F.X., Simonet, J-M. and Decaris, B. Pulsed-field gel electrophoresis analysis of the genome of *Streptomyces ambofaciens* strains. FEMS Microbiol. Letters, 77 (1990) 79–88.

[18] Solenberg, P.J. and Baltz, R.H. Transposition of Tn*5096* and other IS*493* derivatives in *Streptomyces griseofuscus*. J. Bacteriol., 173 (1991) 1096–1104.

[19] Chu, G., Vollrath, D. and Davis, R.W. Separation of large DNA molecules by contour-clamped homogeneous electric fields. Science, 234 (1986) 1582–1585.

[20] Smith, C.L., Matsumoto, T., Niwa, O., Klco, S., Fan, J-B., Yanagida, M. and Cantor, C.R. An electrophoretic karyotype for *Schizosaccharomyces pombe* by pulsed field gel electrophoresis. Nucleic Acids Res., 15 (1987) 4481–4489.

[21] Stam, J.C., Kwakman, J., Meijer, M. and Stuitje, A.R. Efficient isolation of the linear DNA killer plasmid of *Kluyveromyces lactis*: evidence for location and

expression in the cytoplasm and characterization of their terminally bound proteins. Nucleic Acids Res., 14 (1986) 6871–6884.

[22] Hirochika, H. and Sakaguchi, K. Analysis of linear plasmids isolated from *Streptomyces*: Association of protein with the ends of the plasmid DNA. Plasmid, 7 (1982) 59–65.

Chapter 14

DETECTION OF MUTATIONS BY PCR–SINGLE-STRAND CONFORMATION POLYMORPHISM

Amalio Telenti

OUTLINE

Methods in Gene Technology, Volume 2, pages 241–252
Copyright © 1994 JAI Press Ltd
All rights of reproduction in any form reserved.
ISBN: 1-55938-264-3

1. INTRODUCTION

There are a number of techniques available for the detection and characterization of mutations that take advantage of the flexibility and power of PCR (Table 1) [1]. Sequencing remains the gold standard, but it is not very efficient as a screening tool, in particular when evaluating large numbers of samples, or multiple stretches of DNA.

The choice of the most appropriate screening alternative should be guided by the characteristics of the mutation to be identified. A single nucleotide substitution occurring always in the same codon can be approached by mutation (allele)-specific priming. However, those using this approach should be aware that certain 3´-terminal mismatches still allow amplification using *Taq* polymerase [2] and will therefore not discriminate between mutants and wild type strains. An additional mutation may have to be introduced in the primer— usually in the 3rd or 4th positions from the 3´ end—to further destabilise annealing and secure a discriminating amplification [3]. Another simple approach uses restriction enzyme analysis of PCR products to identify restriction sites gained or lost as a result of a mutation. In addition, this approach can be modified to allow for detection of a minority mutant subpopulation—mutant-enriched PCR amplification [4].

When the characteristics of the mutation prevent the application of these simple strategies (e.g. if single or multiple nucleotide substitutions can occur at several possible codons) the other alternatives in Table 1 should be considered. RNase A mismatch analysis involves hybridization of DNA to a labelled wild-type RNA. RNase A will cleave at DNA:RNA mismatches [5]. Such analysis requires the use of RNA, and certain mismatches are resistant to RNase. Constant denaturant and temperature gradient gel

Table 1.   Main PCR-based strategies for detection of mutations.

Sequencing of PCR products
Mutation-specific priming
Restriction enzyme analysis of PCR products
Selective oligonucleotide hybridization
CDGE, constant denaturing gel electrophoresis
TGGE, temperature gradient gel electrophoresis
Ribonuclease A cleavage of mismatched RNA : DNA
    duplexes
SSCP, single strand conformation polymorphism

electrophoresis [6,7] (CDGE and TGGE respectively) rely on dissociation of amplified DNA strands in discrete sequence-dependent melting domains. A theoretical melting map analysis is usually required prior to CDGE and TTGE, and careful control of mutation-specific electrophoresis conditions—concentration of denaturant and temperature—is necessary. In contrast, single-strand conformation polymorphism analysis (SSCP) appears as the most versatile technique, amenable to automation, and optimal for screening of known or unknown mutations.

2. PRINCIPLE

Like RNA, separated strands of DNA adopt a folded conformation as a result of self complementarity and intramolecular interactions. A single nucleotide mutation usually leads to an altered conformation that can be identified as a change in mobility in a non-denaturing gel

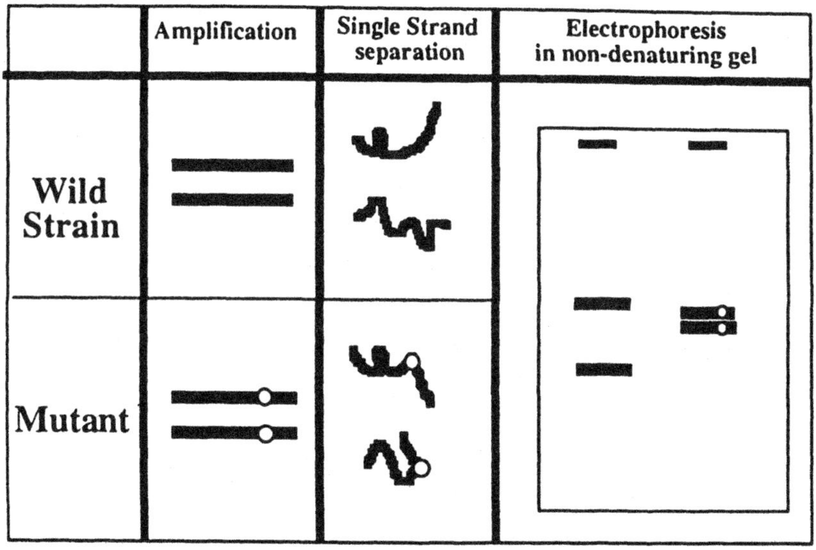

Figure 1. Detection of mutations by PCR-SSCP. After amplification of a fragment from the target locus, single DNA strands are separated by heat denaturation, cooled on ice, and loaded on a non-denaturing polyacrylamide gel. Single nucleotide substitutions will determine an altered folded conformation of the single strands, which will result in a different strand migration pattern.

electrophoresis [8] (Figure 1). To this, which constitutes the basis of SSCP, PCR adds the possibility of precisely selecting and amplifying mutation loci for analysis [9].

The factors that influence a conformation change following a mutation are multiple, and the resulting shift in electrophoretic mobility of separated strands is for the most part unpredictable. Thus, conditions for SSCP have to be established empirically. The following are the main parameters to be considered:

(i) Size of the amplified product: The optimal range for SSCP analysis is 100–300 bp [10]. Larger fragments can be assessed after restriction enzyme digestion to generate optimal-size fragments [11].

(ii) Gel composition: The basic gel is a non-denaturing 5–6% polyacrylamide gel with a low cross-linking profile (1–2% C). The extent of cross-linking (% C) is expressed by the ratio of the percent concentration of bisacrylamide to the concentration of total acrylamide monomer [10]. The addition of 5–10% glycerol may significantly improve gel resolution, perhaps by exerting a weak denaturing action that results on a larger surface of the DNA strand. Some mutations have been identified exclusively in gels containing glycerol. There are also commercially available gel formulations for SSCP analysis, such as the one used in the protocol described below.

(iii) Electrophoresis conditions: Temperature strongly influences the stability of strand conformation and has a significant impact on migration patterns; therefore, it is critical to control this parameter. If possible, gels should be run at low constant power to maintain the gel at room temperature, for example overnight electrophoresis at 6 W for a sequencing size gel [12]. Although this is the simplest approach, other controlled conditions, such as constant cooling of the system to 4°C, or the imposition of a temperature gradient, have been useful in the past [13,14].

The running distance is an additional consideration. Adequate discrimination of many mutations demand the use of sequencing-format gels: the need for more than 20 cm of electrophoresis for good resolution is not unusual. The use of minigels, such as the Pharmacia 'PhastSystem' [15] is only adequate when the differences in conformation polymorphisms are marked and can be resolved in such a short distance.

Using one or two combinations of gel composition and electrophoresis conditions allows detection of most mutations. The protocols described herein, one manual and one automated, using radioactivity or fluorescence respectively, allowed detection of all mutations in the region encoding for the RNA polymerase β-subunit of *Mycobacterium tuberculosis*, that is 20 different nucleotide substitutions in 10 possible codons over a 36 aminoacid region [12,16].

3.   MATERIALS AND METHODS

3.1.   Materials Required

Solutions:
PCR master mix for manual SSCP
  50 mM KCl
  10 mM Tris-HCl (pH 8.3)
  1.5 mM MgCl$_2$
  10% glycerol
  200 μM of dATP, dTTP and dGTP
  100 μM dCTP
  0.5 μL (5 μCi) of α-^{32}P dCTP
  Alternatively, dATP can be partially substituted by α-^{35}S dATP
PCR master mix for automated SSCP:
  50 mM KCl
  10 mM Tris-HCl (pH 8.3)
  1.5 mM MgCl$_2$
  10% glycerol
  200 μM each dNTP
SSCP dilution solution
  10 mM EDTA
  0.1% SDS
Gel loading buffer
  95% formamide
  0.05% Bromophenol Blue
  0.05% Xylene Cyanol
  20 mM EDTA
5 × TBE buffer
  0.45 M Tris-borate
  10 mM EDTA
  pH 8.0

Other reagents
Primers: 0.5 µM of each primer (final concentration). In the case of automated SSCP, both primers must be labelled with fluorescein.

Taq polymerase

α-^{32}P dCTP or α-^{35}S dATP

Acrylamide/bisacrylamide

Mutation detection enhancement gel, MDE (Hydrolink, AT Biochem Inc, Malvern, PA; USA)

TEMED

Ammonium persulphate

Equipment
Sequencing gel equipment and gel dryer for manual SSCP.

Automated SSCP is performed on a Pharmacia sequencer (model ALF).

X-ray film and cassettes for autoradiography.

3.2   Protocol

1. PCR should be optimised (annealing conditions, number of cycles etc.) to generate a single amplification product. As described, manual SSCP requires partial substitution of one dNTP for its radioactive homologoue. For automated SSCP, a conventional PCR mix is used, but primers are labelled with fluorescein.

2. Prepare a non-denaturing gel 0.5× MDE gel composed of 25 mL MDE, 12 mL 5× TBE buffer, and 63 mL H_2O, polymerized with 100 µL of 25% ammonium persulphate and 100 µL of TEMED. Alternatively, a non-denaturing polyacrylamide gel with or without 5–10% glycerol can be used (for 100 mL of gel add 20 mL of acrylamide/bisacrylamide (30:0.8), and 80 mL of 1× TBE).

3. If restriction of the PCR products is desired, cut with the desired enzyme in a volume of 25 µL (10 µL PCR product, 2.5 µL restriction buffer, 0.5 µL (5 units) of the enzyme, and 12 µL water) and proceed as below.

4. Dilute 25 µL of the PCR product in 75 µL of SSCP dilution

solution.

5. Mix 3 µL of the diluted products with 3 µL of gel loading buffer.

6. Denature the samples at 94°C for 10 min, cool on ice, and load 2.5 µL onto the gel.

7. For manual SSCP, electrophoresis is performed overnight at room temperature and constant power (6 W for a 50 × 32 × 0.04 cm gel). Automated SSCP is performed in a Pharmacia sequencer overnight at 25°C and 6 W for a 30 × 35 × 0.05 cm gel.

8. Transfer the manual SSCP gel to filter paper, dry, and expose for 5 h or overnight to X-ray film at –70°C. Automated SSCP results are read directly from the computer screen.

4. COMMENTS

(i) General SSCP strategy: For the detection of well characterized mutations, an initial approach using larger fragments (approximately 1000 bp) and restriction enzymes can help in predicting the size and boundaries of a smaller fragment which will have the best SSCP sensitivity. Based on this information, new primers will be tailored for the specific region. In addition, a restriction site can be used for selectively separating two regions within an area where mutations confer diverse phenotype characteristics. The SSCP will identify which specific region contains a mutation (Figure 2).

Initial screening of large stretches of DNA for polymorphisms or mutations that are not known can be accomplished by the amplification of fragments of approximately 1000 bp. This can be done directly from genomic DNA, or by amplification of cloned material using plasmid-based universal primers (e.g. M13 and reverse M13 primers). Thereafter, using restriction enzymes that cut frequently, the amplified product is empirically fragmented to the optimal size for SSCP (100–300 bp).

Further optimising of detection can be done by using two different gel formulations. It is convenient to run gels at room temperature overnight; and this should be modified only when other strategies have failed.

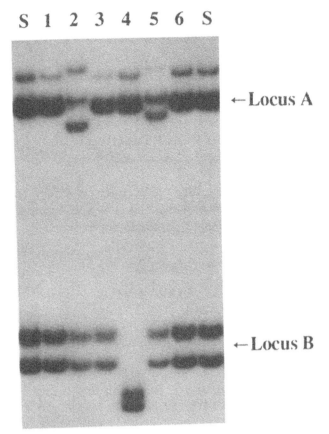

Figure 2.   Selective identification of mutations occurring in two
adjacent regions. A 411 bp fragment of the Mycobacterium
tuberculosis rpoB gene was amplified, cut with HindII, and the
fragments were analysed by SSCP. The restriction enzyme selectively
cut between two areas where mutations confer different drug
resistant phenotypes. Mutations in locus A (lanes 2 and 5) result in a
rifampicin-resistant/rifabutin-sensitive phenotype; mutations in locus
B (lane 4) confer high level resistance to both drugs. Lanes 1, 3, 6,
and reference strain 'S' are sensitive organisms.

(ii) Processivity.   PCR-SSCP is a simple, rapid technique that
easily allows the evaluation of 40 samples per day. Manual SSCP
represents an efficient alternative to manual sequencing, and
automated SSCP offers a substantial time gain and better sample
processivity than automated sequencing (Figure 3).

Manual protocols can be completed in 2 d:

Day 1:

Preparation of sample for PCR by lysis or disintegration (we rarely do a formal DNA extraction)	1 h
PCR	3 h
Product preparation for SSCP and gel casting	1.5 h
Gel loading	0.5 h
TOTAL	6 h

Day 2:

Disassemble	0.5 h
Gel drying	1.5 h
Exposure to X-ray film and film development	5 h
TOTAL	7 h

Automated SSCP using a sequencer will provide data after 24 h.

(iii)   Reproducibility: SSCP patterns are reproducible for each mutation and for each gel and running conditions. With the use of an automated sequencer the observed intra- and inter-test variation in retention times is less than 0.5% [16]. In many cases, the SSCP pattern is characteristic for the mutation and predicts the substitution found by sequencing.

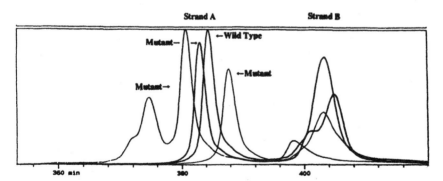

Figure 3.   Automated detection of mutations by PCR-SSCP. After amplification with one of the primers labelled with fluorescein, PCR products were heat-denatured, cooled on ice, and loaded onto a non-denaturing gel fitted in a Pharmacia sequencer. Retention times for the two DNA strands are characteristic and reproducible for each mutation (less than 0.5% intra- and inter-test variation). Shown are 3 mutations of the M. tuberculosis *rpoB* gene and the wild type.

(iv)   Interpretation: Initially, all abnormal SSCP patterns should be evaluated by sequencing to exclude the presence of silent mutations or polymorphisms without biological significance. Later on, only new SSCP patterns need to be sequenced, so that a complete 'library' of mutations can be established.

5.   TROUBLESHOOTING

(i)   Three-band SSCP pattern.   This is the most common artifact found. It likely reflects the existence of two possible conformations for one of the DNA strands, or of limited reannealing to dsDNA. In any case it does not represent a problem for interpretation.

(ii)   Multiple bands.   Usually reflects poor amplification, i.e. the label has been extensively incorporated to non-specific amplification products. The PCR reaction should be optimised (e.g. increasing the annealing temperature) to give a single PCR product on an agarose gel.

(iii)   Smears or unexpected abnormal mobility.   These occur occasionally when working with biological samples with a high content of DNA from different species (e.g. human sputum), or maybe due to the presence of poorly defined interfering factors [16]. They are usually reproducible for a given sample, and have not been resolved by conventional DNA extraction.

(iv)   Failure to detect all mutations.   As described above, some mutations are 'resistant' to SSCP analysis. A comprehensive step-wise approach includes a restriction analysis of the fragment with several enzymes, the use of different gel formulations (commercial MDE gel, polyacrylamide gels with or without 5–10% glycerol), and if these fail, a modification in temperature conditions.

ACKNOWLEDGMENTS

I thank Dr. P. Imboden and Dr. T. Bodmer for critical review of the manuscript. Supported by a grant (32-32406.91) from the Swiss National Science Foundation.

REFERENCES

[1]   van Mansfeld, A.D.M. and Bos, J.L. PCR-based approaches for detection of mutated ras genes. PCR Meth. Appl., 1 (1992) 211–216.

[2]   Kwok, S., Kellog, D.E. and Spasic, D. Effects of primer-template mismatches on the polymerase chain reaction: HIV-1 model studies. Nucleic Acid Res., 18 (1990) 999–1005.

[3]   Jacobson, D.R. and Moskovits, T. Rapid, nonradioactive screening for activating ras oncogene mutations using PCR-primer introduced restriction analysis (PCR-PIRA). PCR Meth. Appl., 1 (1991) 146–148.

[4]   Kahn, S.M., Jiang, T.A., Culbertson, I.B., Williams, G.M., Timita, N. and Ronai, Z. Rapid detection of and sensitive nonradioactive detection of mutant K-ras genes via "enriched" PCR amplification. Oncogene, 6 (1991) 1079–1083.

[5]   Myers, R.M., Larin, Z. and Maniatis, T. Detection of single base substitutions by ribonuclease cleavage at mismatches in RNA-DNA duplexes. Science, 230 (1985) 1242–1246.

[6]   Boerresen, A-L., Hovig, E., Smith-Soerensen, B., et al. Constant denaturant gel electrophoresis as a rapid screening technique for p53 mutations. Proc. Natl. Acad. Sci. USA, 88 (1991) 8405–8409.

[7]   Riesner, D., Steger, G., Wiese, U., Wulfert, M., Heibey, M. and Henco, K. Temperature-gradient gel electrophoresis for the detection of polymorphic DNA and for quantitative polymerase chain reaction. Electrophoresis, 13 (1992) 632–636.

[8]   Orita, M., Iwahana, H., Kanazawa, H., Hayashi, K. and Sekiya, T. Detection of polymorphisms of human DNA by electrophoresis as single-strand conformation polymorphisms. Proc. Natl. Acad. Sci. USA, 86 (1989) 2766–2770.

[9]   Orita, M., Suzuki, Y., Sekiya, T. and Hayashi, K. Rapid and sensitive detection of point mutations and DNA polymorphisms using the polymerase chain reaction. Genomics, 5 (1989) 874–879.

[10]   Hayashi, K. PCR-SSCP: a rapid and sensitive method for detection of mutations in the genomic DNA. PCR Meth. Appl., 1 (1991) 34–38.

[11]   Iwahana, H., Yoshimoto, K. and Itakura, M. Detection of point mutations by SSCP of PCR amplified DNA after endonuclease digestion. BioTechniques, 12 (1992) 64–66.

[12]   Telenti, A., Imboden, P., Marchesi, F., et al. Detection of rifampicin-resistance mutations in Mycobacterium tuberculosis. Lancet, 341 (1993) 647–650.

[13]   Michaud, J., Brody, L.C., Steel, G., et al. Strand-separating conformational polymorphism analysis: efficacy of detection of point mutations in the human ornithine δ-aminotransferase gene. Genomics, 13 (1992) 389–394.

[14]   Imboden, P., Cole, S., Bodmer, T. and Telenti, A. Detection of rifampicin resistance mutations in Mycobacterium tuberculosis and Mycobacterium leprae. In: Persing, D.H., Smith, T. F., Tenover, F.C. and White, T.J. (eds.) *Diagnostic Molecular Microbiology*, American Society for Microbiology, 1993, 6:1.

[15]   Mohabeer, A.J., Hiti, A.L. and Martin, W.J. Non-radioactive single strand conformation polymorphism (SSCP) using the Pharmacia "PhastSystem". Nucleic Acid Res., 19 (1991) 3154.

[16] Telenti, A., Imboden, P., Marchesi, F., Schmidheini, T. and Bodmer, T. Automated detection of rifampicin-resistant Mycobacterium tuberculosis by PCR and single strand conformation polymorphism analysis. Antimicrob. Agents Chemother., 37 (1993) 2054–2058.

Chapter 15

ANALYSIS OF DNA SUPERCOILING

Karl Drlica, Muhammad Malik, Jian-Ying Wang, Andrzej Sasiak, and Richard R. Sinden

OUTLINE

Methods in Gene Technology, Volume 2, pages 253–280
Copyright © 1994 JAI Press Ltd
All rights of reproduction in any form reserved.
ISBN: 1-55938-264-3

1.   INTRODUCTION

DNA is said to be supercoiled when it has either an excess or a deficiency of duplex turns relative to a nicked or linear DNA with the same number of base pairs. An excess of turns causes positive supercoiling, and a deficiency of turns, which is the usual case for DNA extracted from cells, results in negative supercoiling. This global property of circular DNA is partitioned into writhe and a change in twist. Intuitively, writhe can be thought of as coiling, much like that seen in electron micrographs of supercoiled DNA [1]. Writhe probably influences activities that involve DNA looping and wrapping of DNA around proteins. Change in twist represents an altered winding of the helix. This local duplex deformation probably influences the energetics of strand separation and the rotational orientation between nearby protein binding sites (see, for example, [2]).

Inside cells, DNA is additionally partitioned into restrained and nonrestrained supercoils. This was first revealed for bacteria when the introduction of nicks, which *in vitro* would completely relax DNA, relieved only half of the intracellular supercoils [3]. Nicks relaxed an even smaller fraction of supercoils in eukaryotic cells [4]. There most of the restraint probably arises from the wrapping of DNA into nucleosomes by histones; the factors responsible for restraint of supercoils in bacteria have not been identified. As a result of restraining factors, intracellular supercoils, which we refer to as superhelical tension, represent only a fraction of the total supercoiling (linking deficit) measured with extracted DNA. Still other restraints partition DNA into topologically independent domains [5,6]. Such restraints allow DNA species known to be linear, such as bacteriophage T4 DNA, to be under superhelical tension [7].

The physiological importance of supercoiling began to emerge with the discovery of DNA topoisomerases, enzymes that alter levels of supercoiling. In bacteria, DNA gyrase [8] and DNA topoisomerase I [9] control levels of supercoiling through their opposing activities (for review see [10]). This was illustrated by a genetic study in which mutation of topoisomerase I, which leads to abnormally high levels of negative supercoiling [11], was compensated by mutations of gyrase that lowered negative supercoiling [11,12]. Subsequent studies suggested that cellular energetics, translocational processes such as transcription, and supercoiling itself play a role in the control of supercoiling [10].

Supercoiling influences the activities of DNA in several ways. First, negative supercoiling reduces the energy needed for strand separation and wrapping of DNA into specific DNA-protein complexes. Second, the change of helical pitch is expected to influence the binding of proteins to DNA through orientation of binding sites (for examples of promoter effects, see [2]). Third, writhe may facilitate DNA looping associated with gene regulation (for review, see [13]). To define these effects for particular systems *in vivo*, supercoiling has been altered by perturbing topoisomerase activities with mutations or specific inhibitors. The interpretation of each of these studies depends on measurement of supercoiling.

In the following sections we discuss four types of measurement: (1) supercoiling of extracted plasmid DNA using partial dye titration coupled with gel electrophoresis; (2) supercoiling of extracted chromosomal DNA using dye titration with density-gradient centrifugation; (3) intracellular superhelical tension of plasmid and chromosomal DNA using a combination of torsionally-tuned conformation probes and Me_3-psoralen crosslinking; and (4) intracellular superhelical tension detected by binding of radioactive Me_3-psoralen. The first two methods provide evidence that particular cellular perturbations alter topoisomerase activity (for examples, see [10]). However, they do not distinguish among changes in twist, writhe, restrained, or nonrestrained supercoiling. The last two methods, which measure intracellular superhelical tension, support conclusions about topoisomerase activities derived from linking deficit determinations and further establish that a major portion of linking deficit is restrained (for review see [14]). They also show that tension can vary from one region to another [15,16]. Readers interested in measurement of intracellular superhelical tension using a combination of restriction endonucleases and Me_3-psoralen crosslinking are referred to [17–19].

2. SUPERCOILING OF ISOLATED PLASMIDS

2.1 Introduction

When plasmids are nicked and ligated *in vitro*, their linking number assumes a Boltzmann distribution due to thermal fluctuation at the time of ligation [20,21]. Most of the population falls within 5 or 6 topoisomer bands during gel electrophoresis, with one or two bands being most heavily populated [22]. A slightly broader distribution is generally seen for plasmid DNA extracted from cells [22]. Adjacent bands differ by one topological turn (one supercoil; [20,21–23]). In principle, the mean linking deficit for a supercoiled plasmid population $(L - L_o)$ can be determined by electrophoresis of the supercoiled plasmid population and its relaxed counterpart in neighbouring lanes of an agarose gel (*in vitro* relaxation can be achieved by nicking-ligation or by relaxation with eukaryotic topoisomerase I). To estimate $L - L_o$ one determines the centres of the topoisomer distributions (see [20]) for the supercoiled and relaxed populations and then measures the centre-to-centre difference in electrophoretic mobility in terms of topoisomer bands. If the two distributions overlap, determining this difference by band counting is straightforward [22]. If the two populations do not overlap, it is necessary to prepare a population of topoisomers that spans the difference. This preparation, which is generated by nicking and ligating under the appropriate conditions [22], is then placed in another neighbouring lane of the gel where it can be used for band counting.

Since it is difficult to obtain a plasmid relaxed under intracellular conditions, physiological values of $L - L_o$ are generally not determined. Instead, it has become customary to compare electrophoretic mobilities of plasmids isolated from cells treated in different ways. The difference in mobility, in terms of topoisomer bands, estimates $(L_{treat\ 1} - L_{treat\ 2})$, which in turn reflects changes in the net activities of the topoisomerases. However, the measurement of interest is the percentage change in linking deficit, since the absolute effect on linking number $(L_{treat\ 1} - L_{treat\ 2})$ varies with plasmid size. Calculation of percentage change requires an estimate of $L_{treat\ 1} - L_o$, which is usually approximated by comparing plasmid relaxed *in vitro* with plasmid relaxed *in vivo*. For a plasmid such as pBR322 isolated from *E. coli* $(L_{treat\ 1} - L_o)_{in\ vitro}$ is on the order of –30 [ref. 11]. Thus, when a treatment causes a three-band change in the mean linking number for pBR322, the percentage change is about 10%.

To display populations of topoisomers, gel electrophoresis is carried out in the presence of an intercalating dye at a concentration that partially titrates the supercoils. If a series of gels is run in which the dye concentration is varied, negatively supercoiled DNA will gradually migrate more slowly as the dye concentration is increased and the supercoils are titrated. A minimum will be obtained where the average linking number is the same as nicked DNA. With further increases in dye concentration the topoisomers become positively supercoiled. As a result, they will migrate more rapidly in the gel with increasing dye concentration. For a given experiment, it is *essential* to know whether a particular dye concentration is generating negative or positive supercoils in the gel, since the interpretation of the differences among samples will be very different. This can be determined by carrying out the electrophoresis at several dye concentrations.

If the population of topoisomers is highly heterogeneous, gel electrophoresis can be performed in two dimensions to display most of the members of the population [24]. The second dimension generally contains a higher concentration of chloroquine than the first. This allows the gel to distinguish topoisomers that are: (1) negatively supercoiled in both dimensions; (2) negatively supercoiled in the first but positively supercoiled in the second; and (3) positively supercoiled in both dimensions. This method has proven especially useful in the analysis of the effects of *top*A mutants on supercoiling [24,25], and it can be used to determine the linking deficit required to convert a potentially Z-DNA-forming region from B to Z-form DNA.

There are several aspects of plasmid supercoiling measurement that require care. First, cell growth conditions are important because supercoiling varies with medium, temperature, aerobic conditions, and growth phase [10]. Second, how the plasmids are recovered from cells may be important, since topoisomerase activity during harvesting of cells can confound the interpretation of an experiment. We generally cool cells very rapidly (with some gyrase mutants gradual cooling allowed enough renaturation of gyrase to markedly change the supercoiling measurement). A case has been found in which cooling prior to cell lysis results in different values for supercoiling than lysis at the temperature of cell growth [26]. This makes special harvesting precautions necessary for certain experiments. A third factor is transcription [25,27]. Transcriptional effects can be minimized by using a plasmid such as pUC9, which lacks the tetracycline-resistance gene (membrane binding of Tet is thought to anchor the transcription-translation apparatus and in special situations lead to high levels of supercoiling (for review see [28]). Rifampicin treatment can be used to confirm that supercoiling is unaffected by transcription (see [29]). A fourth issue concerns the use of plasmids as reporters for changes in chromosomal supercoiling. Changes of plasmid linking number, caused either by a shift in growth condition or a perturbation of topoisomerase activity, generally correlate with linking number change of chromosomes [11,30]. However, plasmid supercoiling does not always parallel chromosomal supercoiling [29]; thus, both types of measurement are necessary to establish that an effect is general.

2.2   Isolation of Plasmids

2.2.1   Buffers and Reagents

Glucose-Tris-EDTA
   25 mM Tris-HCl pH 8
   10 mM Na$_2$EDTA
   50 mM glucose
Alkaline SDS
   0.2 M NaOH
   1% SDS
   prepare immediately before use
5 M potassium acetate, pH 4.8

TE buffer
  10 mM Tris-HCl pH 8, 1 mM Na₂EDTA
PCI
  phenol : chloroform : isoamyl alcohol [24 : 24 : 1]

2.2.2 Procedure

Among the many isolation procedures that have proved satisfactory, we prefer an alkaline lysis method derived from [31,32] (see also Chapters 4–6). In the following protocol, the volumes indicated are suitable for about 10^9 cells.

1. Cells are harvested by centrifugation, resuspended in 0.1 mL of glucose-Tris-EDTA, and incubated at room temperature for 5 min.
2. Then 0.2 mL of alkaline SDS (freshly prepared) is added with gentle mixing, and the suspension is incubated on ice for 5 min.
3. To this sample is added 0.15 mL precooled 5 M potassium acetate, pH 4.8, followed by mixing and incubation on ice for 5 min.
4. The sample is centrifuged at 30 000 g for 5 min at 4°C to remove chromosomal DNA and cellular debris. At this point, the plasmid DNA can be quickly purified with a commercially available column such as the Plasmid Midi Kit from Qiagen (see Chapter 5).
5. If column purification is not used, then the supernatant fluid is removed, mixed with an equal volume of isopropanol, and allowed to stand at room temperature for 15 min to precipitate the nucleic acids. Precipitates are collected by centrifugation (15 000 g for 5 min), and the pelleted material is washed with 70% ethanol at room temperature. The alcohol is removed, and the sample is dried under vacuum.
6. To remove RNA, the pellet is resuspended in 0.1 mL of TE, and then it is incubated with boiled pancreatic RNase (5 µg) at 37°C for 15 min. To remove protein, the DNA solution is extracted with PCI. After centrifugation, the aqueous phase is again extracted with PCI, followed by extraction with an equal volume of chloroform:isoamyl alcohol (24 : 1).
7. The plasmid DNA is then precipitated by adding 1/10 volume of either 3 M sodium acetate or 7.5 M ammonium acetate and 2.5 volumes of absolute ethanol for 15 min. Precipitated DNA

is collected by centrifugation, and DNA-containing pellets are washed with 70% ethanol and dried under vacuum. The DNA is resuspended in 30 μl of TE.

2.3   Gel Electrophoresis

2.3.1   Buffers and Reagents

10 × TPE Buffer [33]
Tris base	108 g
85% phosphoric acid	15.1 mL
0.5 M Na₂EDTA pH 8.0	40 mL

Ethidium bromide
   2 μg/mL in 1 mM Na₂EDTA

2.3.2   Procedure

Electrophoresis is carried out in horizontal slab gels composed of 1% agarose prepared in a buffer such as TPE [33] containing various concentrations (5–25 μg/mL) of chloroquine [24]. It is best to add the dye after the agarose solution has been boiled [24]. Samples (containing about 1 μg of DNA) and indicator dyes are loaded into wells of the gel, and electrophoresis is conducted at a low voltage (1–3 V/cm) with recycling of chloroquine-containing running buffer. In some cases sharper bands can be obtained by performing electrophoresis at 4°C. After electrophoresis has been completed, the gel is soaked with several changes in distilled water to remove the chloroquine. Addition of 1 mM MgSO₄ to one of the soaking steps improves removal of the dye. The gel is then stained with 2 μg/mL of ethidium bromide in 1 mM Na₂EDTA for several hours. Gels are destained about 1 h in 1 mM MgSO₄ before being photographed with ultraviolet light using Polaroid type 667 film. For publication purposes, photographs of the gels should be slightly overexposed.

For two-dimensional gels [24], DNA is electrophoresed in one dimension as described above. Then the top of the gel, which contains the loading wells, is removed and discarded. The remainder of the gel is then soaked for about 5 h at 4°C at a chloroquine concentration 3–5 times that used in the first dimension. The gel is rotated 90° relative to the first dimension electrophoresis, and second dimension electrophoresis is performed as described above. When two-dimensional electrophoresis is used to detect secondary structure transitions, the agarose concentration is generally higher (1.75%) and recycling of the running buffer becomes especially important.

3.   CHROMOSOMAL DNA SUPERCOILING

3.1   Introduction

Average linking deficit of chromosomal DNA of *Escherichia coli* K-12 can be compared among bacterial strains and physiological treatments by extracting nucleoids and then titrating the negative supercoils with ethidium bromide. As the ethidium concentration increases, nucleoid sedimentation rate decreases until a minimum is reached. At that point, which is called the critical dye concentration, all of the supercoils have been removed. At higher dye concentrations, positive supercoils are introduced by the dye. Sedimentation rate then increases. The dye concentration at the sedimentation minimum is related to superhelix density. This general strategy has been used to show that inhibition of DNA gyrase lowers the dye concentration at the minimum, a result interpreted as a relaxation of supercoils or linking deficit [34–36]. A point mutation in *topA* [11], treatment with inhibitors of gyrase at low concentration [35,37] or a shift to anaerobic conditions [30] raise the dye concentration at the sedimentation minimum. This is the result expected for increases in supercoiling.

3.2   Sedimentation Measurements

3.2.1   Buffers and Reagents

Tris-sucrose-NaCl
  0.01 M Tris-HCl pH 8.1
  20% sucrose
  0.1 M NaCl
Lysozyme
  4 mg/mL freshly prepared solution in 0.1 M Tris-HCl pH 8.1,
  0.05 M Na$_2$EDTA
NaCl-deoxycholate
  2 M NaCl
  0.01 M Na$_2$EDTA
  0.4% sodium deoxycholate

1% polyoxyethylene 20 cetyl ether (Brij 58)

Scintillation fluid
    4 g 2,5 diphenyloxazole (PPO)
    0.2 g 1.4-bis[4-methyl-5-phenyl-2-oxazolyl]benzene (dimethyl
    POPOP) dissolve in 1 L of toluene and add to 500 mL Triton X-100
    and 200 mL water.

TCA
    10% Trichloroacetic acid

3.2.2 Procedure

1. Nucleoid sedimentation rate is generally measured by
    centrifugation in sucrose density-gradients. Operationally, cells
    are first grown in liquid medium (LB or M9) in the presence of
    ^{3}H-thymidine ($10 \mu Ci/mL$) for about 0.5 generations to
    radioactively label chromosomal DNA.
2. Cells (4 mLs grown to midlog phase) are then rapidly chilled,
    concentrated by low-speed centrifugation, and resuspended in
    0.1 mL of Tris-sucrose-NaCl on ice.
3. A brief (often less than 60 s) lysozyme treatment is then carried
    out on ice with egg white lysozyme (25 μL).
4. Then the suspension is diluted in half by addition of NaCl-
    deoxycholate and 1% polyoxyethylene 20 cetyl ether (Brij 58)
    followed by incubation at 20–25°C until turbidity disappears.
    For some strains it may be necessary to add 0.5% Sarkosyl to
    the detergent solution for effective lysis.
5. Aliquots are immediately chilled and loaded onto a series of
    linear 10–30% (w/v) sucrose density-gradients containing 1 M
    NaCl, 0.025 M Tris-HCl pH 8, 0.01 M Na$_2$EDTA and ethidium
    bromide ranging from 0 to 4 μg/mL. The high concentration of
    NaCl is required to keep the nucleoids in a compact
    configuration. On each sucrose gradient is also layered a small
    aliquot of [^{14}C]-labelled bacteriophage T4B, which serves as a
    sedimentation marker ($S^{T4} = 1025S$ [38]). The samples are then
    centrifuged for about 30 min at 17 000 r.p.m. (Beckman
    SW50.1 rotor or the equivalent) at 4°C.
6. Gradients are fractionated from the bottom of each centrifuge
    tube, and radioactivity in each fraction is measured. The
    average chromosome sedimentation coefficient is determined

by comparing its distance sedimented relative to the distance sedimented by the marker phage. These distances are determined from plots showing the distribution of radioactivity in each density-gradient. In most situations it is adequate to assume that sedimentation rate is linear and drop size is uniform (this may not be true if lysate volumes are large, since they contain detergents that change drop size).

Care and experience are required to obtain cell lysates in which the nucleoid remains compact and the DNA-containing solution is nonviscous. Incubation times for the steps in the cell lysis procedure are important and need to be optimized for each bacterial strain. For *E. coli* K-12, suitable conditions often produce cell lysis (loss of turbidity) after 5 min incubation at 20–25°C. More rapid clearing is undesirable because DNA aggregation often occurs. Incubation in the detergent solution for more than 10 min also tends to give lower yields of nucleoids. Once cell lysis has occurred, small aliquots (25 μL) are transferred to prechilled sucrose gradients using large bore pipettes to avoid shearing forces that tend to break the DNA. DNA breakage is also minimized by preparing and loading the gradients in dim light. Carrying out all cell lysis and gradient loading operations rapidly is important for obtaining satisfactory yields (from the time lysates are chilled to the time ultracentrifugation begins should be no more than 10 min).

In most cases, sucrose gradients can be fractionated directly into vials for determination of radioactivity using a suitable scintillation fluid. There are times, however, when unincorporated radioactive thymidine at the top of a gradient will mask the nucleoids. In such cases, gradients can be fractionated onto paper filters or filter paper strips. DNA is then precipitated by treatment of filters with 10% TCA followed by two washings in 1 N HCl, one in water, and two in 95% ethanol. These washings are carried out at 4°C and can be done batchwise (filters should be numbered with pencil). After drying, the radioactivity on the filters can be determined in a scintillation cocktail lacking water and Triton X-100.

Visual inspection of radioactivity profiles is adequate for determining sedimentation rates, but the quality of the data is sensitive to the amount of care taken in preparing ethidium solutions (checking concentrations spectrophotometrically) and in plotting the radioactive distributions (taking into consideration partial fractions at the top of the gradients).

3.3   Quantitative Interpretation

The dye concentration required to obtain the sedimentation minimum is determined by finding the centre of the trough of the titration curve. This value is related to the critical dye concentration and therefore to superhelix density through the Scatchard relationship as described in [39]. In this relationship,

$$\frac{r}{C_f} = K(n - r) \tag{1}$$

where r is the molar ratio of bound dye per nucleotide, C_f is the molar concentration of free dye, K is the intrinsic association constant in litres per mole, and n is the maximum number of binding sites per nucleotide.

For 1 M NaCl, the value of K is taken as 0.98×10^5 L/mol and the value of n as 0.18 [39,40]. For situations where DNA concentration is low, as it is in the sucrose density-gradient analyses usually performed, we assume that the free dye concentration, C_f, equals the total dye concentration at the sedimentation minimum. Thus we solve for r and then determine its fractional change due to the particular physiological treatment under consideration.

It is important to stress that the numbers represent average values with respect to both the chromosomal population and the topologically independent domains of the chromosome [5,6]. Broadening of the titration trough indicates an increase in heterogeneity, a feature which has been observed with plasmid DNA [29].

4.   TORSIONALLY TUNED Z-DNA PROBES TO MEASURE LOCAL SUPERCOILING IN VIVO

4.1   Introduction

Z-DNA (and cruciform)-forming sequences, when used with 4,5′,8-trimethylpsoralen (Me$_3$-psoralen)-based assays that detect conformational differences, constitute 'torsionally tuned probes' for measuring the level of DNA superhelical tension at specific sites in plasmid, bacterial, and eukaryotic chromosomes [41,42]. These assays are based on the observation that for any particular alternating purine-pyrimidine sequence, the fraction of molecules that adopt a Z-conformation depends on superhelical density. That fraction, when assessed by Me_3-psoralen binding *in vivo*, reflects the level of unrestrained superhelical tension. A set of torsionally tuned Z-DNA probes is now available for estimating unrestrained tension under a variety of conditions [42], and a sensitive exonuclease III (exo III)/photoreversal assay has been developed for measuring Me_3-psoralen binding in living cells [43,44]. ExoIII, which digests a single strand of double-stranded DNA from a recessed 3′ OH end, stops quantitatively 2 bases on the 3′ side of a base containing a photobound Me_3-psoralen molecule, but it does not stop at Me_3-psoralen monoadducts in the nondigested strand [44]. Thus, the site and frequency of photoaddition can be determined with base-pair resolution by digesting a crosslinked-sample of DNA with exoIII and then analyzing the digestion products on a DNA sequencing gel. This assay can be utilized at levels of Me_3-psoralen photobinding as low as 1 crosslink per 8000 bp [43,44].

The supercoiling analysis involves Z-DNA-forming sequences that have a photoreactive 5′ TA dinucleotide within the Z-DNA-forming region (Z–TA) and other photoreactive sites at the B–Z junctions (B–ZJ). The rates of photobinding of Me_3-psoralen to reactive sites, assayed by the intensity of exoIII stops, are determined within Z-DNA regions and at the 5′ AATT of the B–Z junctions relative to the rate of photobinding at a control 5′ TA (C–TA) outside the Z-DNA-forming region. Two ratios are calculated, Z–TA/C–TA and B–ZJ/C–TA, that are diagnostic for the conformation of the DNA. To generate standard curves for determining superhelical densities from relative rates of photobinding, topoisomer populations are examined by two-dimensional gel electrophoresis. This establishes the relationship between the relative reactivities of the Z-DNA region to Me_3-psoralen photobinding and the fraction of individual topoisomers containing Z-DNA in the DNA population [43,44]. Analysis of the reactivity of Me_3-psoralen *in vivo*, when calibrated by the standard curve, then provides an estimate of the effective superhelical density of the intracellular DNA.

4.2 Me$_3$-psoralen Photobinding to DNA in Bacterial Cells

4.2.1 Buffers and Reagents

M9 buffer
  1 g NH$_4$Cl
  5.8 g Na$_2$HPO$_4$
  3.0 g KH$_2$PO$_4$
  dissolve in 1 L of distilled water
K medium
  10 g casamino acids
  10 g glucose
  0.03 g MgCl$_2$
  0.07 g CaCl$_2$
  0.01 g thiamine
  in 1 L of M9 buffer
Me$_3$-psoralen solution
  saturated Me$_3$-psoralen solution in ethanol

4.2.2 Procedure

1. Bacterial cells are grown to a density of 4×10^8 cells/mL in K medium. The cells are then harvested by centrifugation and resuspended at a density of about 2×10^9 cells/mL in M9 buffer at 4°C to prevent the introduction of uvrABC-dependent nicks at DNA monoadducts and crosslinks (subsequent steps are carried out in a 4°C cold room. If cells are grown in Luria broth, they must be washed once in cold M9 buffer before resuspending for the Me$_3$-psoralen treatment).

2. Immediately after resuspension, Me$_3$-psoralen is added to saturation (10 µL of a saturated Me$_3$-psoralen solution in ethanol per mL of cell suspension to create a 1% v/v solution).

3. After incubation on ice for 2–5 min to allow equilibration with the Me$_3$-psoralen, the cells are transferred to a plastic petri dish (30–150 mm), such that the layer of cells is as thin as possible (usually about 2–4 mm).

4. Cells are then irradiated with 360 nm light using a standard fluorescent light fixture with General Electric BLB blacklight bulbs over a dose range of 0–10 kJ/m².

    If necessary, higher incident light intensities can be obtained with a Blak-Ray B-100A ultraviolet lamp with either a spot or flood ultraviolet bulb (100 W). To ensure uniform irradiation, the cell suspension should be mixed occasionally. Since Me_3-psoralen photobinding drives Z-DNA into the B-DNA conformation, the reactivity of psoralen photobinding varies as a function of the light dose. Consequently, it is necessary to extrapolate Me_3-psoralen photobinding to zero time, which usually requires analysis of 4–5 different irradiation times for each cell sample analyzed [43,44].

5. Following irradiation, supercoiled DNA is purified using a cleared lysate procedure [4; see also Chapter 4]. Since Me_3-psoralen photobinds readily to the 5′ TA in the Z-DNA-forming sequence in nicked DNA, which will affect the analysis of the fraction of sample existing as Z-DNA, it is important to remove nicked DNA. This is done by equilibrium centrifugation in CsCl-ethidium bromide [4] (see Chapter 4).

4.3 Exonuclease Assays for Monoadducts and/or Crosslinks in Bacterial DNA

4.3.1 Background

Exonuclease III assays involve four steps: (1) purification of the DNA fragment of interest radiolabelled at a unique 5′ end; (2) digestion to completion with exoIII; (3) photoreversal of DNA crosslinks; and (4) analysis of the digestion products by electrophoresis into a denaturing 5% polyacrylamide gel. Although exoIII assays are suitable for the analysis of torsionally tuned probes cloned into any small plasmid DNA, the procedure described below applies specifically to pUC8 in which Z-DNA-forming sequences are inserted into the polylinker region. For pUC8, digestion with *Pvu*II, followed by digestion with an enzyme that cuts within the polylinker on the 3′ side of the Z-DNA insert, generates the linear fragment required for analysis.

There are two general points to consider when designing this protocol for a specific application. First, when using a DNA labelled at a blunt-end and also containing a recessed 3′ OH end to direct exoIII digestion to the end opposite to the label, it is best to have the Z-DNA-forming region nearest the recessed 3′ OH. Second, some restriction endonucleases are more sensitive than others to inhibition by Me_3-psoralen adducts or crosslinks at or near the restriction site.

4.3.2   Buffers and Reagents

Phosphatase buffer
  50 mM Tris-HCl pH 9.0
  1 mM $MgCl_2$
  0.1 mM $ZnCl_2$
  1 mM spermidine
10 × STE buffer
  100 mM Tris-HCl pH 8.0
  1.0 M NaCl
  10 mM Na_2EDTA
10% sodium dodecyl sulphate
T4 kinase buffer
  50 mM Tris-HCl pH 9.5
  10 mM $MgCl_2$
  5 mM dithiothreitol
  5% glycerol
BamHI buffer
  50 mM Tris pH 8.0
  10 mM $MgCl_2$
  100 mM NaCl
TB buffer
  40 mM Tris-borate, pH 8.3
exonuclease buffer
  50 mM Tris-HCl pH 8.0
  5 mM $MgCl_2$
  10 mM beta-mercaptoethanol
TBE buffer
  100 mM Tris-borate, pH 8.3
  1 mM Na_2EDTA

4.3.3   Procedure

Labelling at unique 5′ end: The first step is to purify a double-stranded DNA fragment having a radioactively labelled 5′ blunt end and a recessed 3′ OH at the other end.

1. DNA (5 to 15 µg) is suspended in 100 µl of *Pvu*II buffer with 30 units of *Pvu*II and 0.5 unit of calf intestinal alkaline phosphatase (CIAP, Molecular Biology Grade, Boehringer

Mannheim) followed by incubation at 37° for 8–20 h.

2. The digested sample is extracted three times with phenol and three times with chloroform/isoamyl alcohol (24 : 1), adjusted to 0.3 M sodium acetate, and precipitated by the addition of two volumes of 100% ethanol.

3. This precipitate is collected by centrifugation, and the resulting pellet of DNA is rinsed with 70% v/v ethanol and dried.

An alternative procedure is to omit CIAP from step 1 above and dephosphorylate the DNA separately (separate dephosphorylation may better remove the phosphate from the 5′ end). For this, follow steps 1–3 (omitting CIAP), then follow steps 4–7 below; otherwise proceed directly to step 8.

4. The DNA is first resuspended in 50 μL of phosphatase buffer with 1 unit of CIAP. Then the samples are incubated for 15 min at 37°C and for 15 min at 56°C.

5. An additional unit of CIAP is added, and the sample is incubated for another 15 min each at 37°C and 56°C.

6. Next, 40 μL of 10 × STE buffer and 5 μL of 10% sodium dodecyl sulphate are added followed by incubation for 15 min at 75°C.

7. The DNA sample is then extracted with phenol and chloroform /isoamyl alcohol (24 : 1), adjusted to 0.3 M sodium acetate, and precipitated by the addition of two volumes of 100% ethanol. The precipitate is collected by centrifugation for 10 min at 15000 g in a microcentrifuge.

8. For kinase treatment, the sample (from step 3 or 7) is dissolved in 20 μL of T4 kinase buffer. T4 kinase (15 units) and 10 μCi of ^{32}P-ATP is added followed by incubation at 37°C for 30–45 min. This reaction is stopped with Na$_2$EDTA (20 mM final concentration), and DNA is precipitated with ethanol as above.

To produce a recessed 3′ OH, one of several restriction endonucleases can be used. For Z-DNA-forming sequences inserted at the *Eco*RI site of pUC8, *Bam*HI works well.

9. The DNA from step 8 is resuspended in 100 μL of *Bam*HI buffer, 40 units of *Bam*HI is added, and the sample is incubated at 37°C for 4 to 20 h.

10. After extraction with phenol and chloroform and precipitation with ethanol as described above, the sample is dissolved in 20 μL of TB buffer and further purified by electrophoresis (300 V, 15 mA for 5 h) into a 5% polyacrylamide gel (40 × 30 ×

0.15 cm) in TB buffer containing 10% v/v glycerol. The location of the fragment of interest, in this case the 273 bp *Pvu*II-*Bam*HI fragment, is determined by autoradiography, and that region of the gel is excised.

11. DNA is eluted from the excised portion of the gel by soaking the masticated gel slice in 300 μL of 1 M sodium acetate, 2 mM Na$_2$EDTA, and 10 μg tRNA at 37°C for at least 2 h.

12. The solution is next passed through a siliconized glass-wool pad in the bottom of a 1.5 mL microfuge tube by centrifugation at 300–500 g for 30 s. It is then clarified by centrifugation at 15 000 g for 15 min.

13. DNA is precipitated from the supernatant fluid with ethanol, and the resulting DNA pellet is rinsed with 70% ethanol and lyophilized to dryness.

Digestion with exonuclease III
1. The DNA sample is suspended in 100 μL of exonuclease buffer with 200 units of exoIII followed by incubation at 37°C for 10–20 h.

2. This sample is extracted twice with phenol:chloroform (1:1), adjusted to 0.3 M sodium acetate, precipitated by addition of 400 μL of 100% ethanol, rinsed with 70% ethanol, and dried under vacuum.

3. Digestion by exoIII is then repeated as in Step 1 above.

Photoreversal
1. The DNA sample (<20 μg) is dissolved in 50 μL of distilled water and placed in the cap of an inverted 1.5 mL Eppendorf tube. Then it is irradiated for 10 min with a 254-nm (germicidal) lamp at a distance of 15 cm at room temperature (light intensity at about 0.54 kJ m^{-2} min^{-1} or 900 μW cm^{-2}).

2. Glycogen (2 μg) is added, and the sample is then precipitated by the addition of 6 μL of 3 M potassium acetate and 200 μL of 100% ethanol with incubation at –20°C for 2–3 h.

Electrophoretic analysis. The fourth step is electrophoretic analysis with an 80 × 30 × 0.4 cm 5% polyacrylamide gel containing 8 M urea in TBE buffer.

1. The sample is dissolved in deionized formamide to about 2 × 10^3 cts/min per 3 μL portion and then applied to the gel. As a

nucleotide sequence marker, the *Pvu*II-*Bam*HI 5′ end-labelled, gel-purified fragment is prepared by the Maxam–Gilbert or the dideoxy sequencing method [45,46]. Electrophoresis is carried out at 300 V, 60 mA for 3–6 h.

2. The gel is transferred to Whatman 3MM filter paper, dried, and autoradiographed using Kodak XAR-5 film. Densitometric analysis of the autoradiograms or quantitation using a Molecular Dynamics photoimager is generally satisfactory to determine values for Z–TA/C–TA and B–ZJ/C–TA.

4.4 Primer Extension Assay for Eukaryotic or Prokaryotic Genomes

4.4.1 Background

While the exoIII assay described above is suitable for plasmids, more sensitivity is required to measure Me_3-psoralen reactivity with eukaryotic or bacterial genomes. The APEX technique developed by Cartwright and Kelly [47] results in the linear amplification of a primer extension product, and this general method has been used to map UV-induced photoproducts in yeast genes [48]. Below we describe a Taq polymerase primer extension assay to map Me_3-psoralen photoproducts in plasmid DNA containing a region of potential Z-DNA that is more sensitive and much faster than the exoIII/photoreversal assay described in the previous section. Using published standard curves [49], this method will supercede the exoIII strategy for investigators with access to PCR machines.

4.4.2 Buffers and Reagents

Stop buffer
    0.375 M sodium acetate
    2.5 mM Na_2EDTA
primer extension reaction buffer
    10 mM Tris-HCl pH 8.3
    10 mM KC1
    4 mM $MgCl_2$

4.4.3 Preparation of defined fragment

After Me_3-psoralen photobinding is performed, as described above, digestion with a restriction endonuclease is carried out to generate a defined fragment for probing. In the case of pUC8-based plasmids, 100 ng of DNA is sufficient for a good signal, and for 1–5 µg of DNA a 400 bp PvuII fragment is suitable. (This step may not be necessary; however, use of a defined fragment allows observation of the full length primer extension product, which demonstrates that the primer extension reaction is working efficiently.) After digestion with PvuII, the DNA is extracted once with phenol, once with a chloroform : isoamyl alcohol mixture (24 : 1 v/v), adjusted to 0.3 M sodium acetate, and then precipitated by the addition of two volumes of 100% ethanol.

4.4.4   End-labelling of primers

For the primer extension reaction, 20 bp primers, designed to hybridize 100–150 bp from the site of interest, have proven satisfactory. The primer (1 µg) is end-labelled with 20 µg of γ-^{32}P-ATP (6000 Ci/mmol) in 20 µL reaction volume as described above. The reaction is terminated with 80 µL of stop buffer, and the DNA is precipitated with 4 volumes of 100% ethanol. A second ethanol precipitation is performed to ensure removal of unincorporated ATP.

4.4.5   Primer extension

1. For primer extension of Me$_3$-psoralen photoproducts, the DNA sample (100 ng) is mixed with 20 ng labelled primer in primer extension reaction buffer plus 0.5 mM of each dNTP and 5 units of Stoffel fragment polymerase (Cetus-Perkin Elmer; final volume 50 µL).
2. The samples are overlayed with mineral oil, and primer extension reactions are performed in a thermal cycling unit. Ten to twenty cycles are generally performed with additional enzyme added after the first ten. For the Z-DNA forming probe sequence (CG)$_6$TA(CG)$_2$(TG)$_8$, denaturation is at 96°C for 2 min, hybridization is at 60°C for 30 s, and elongation is at 80°C for 5 min. This high temperature is required to replicate through the G + C rich Z-DNA-forming region (a lower temperature may be adequate for a (GT)$_n$ Z-DNA-forming sequence).
3. When cycling is complete, the oil is removed, 50 µL of water is added to increase the volume, and the sample is extracted with

50 µL of a phenol : chloroform mixture (50 : 50 v/v).
4. The sample is adjusted to 0.3 M sodium acetate, 5 µg of glycogen is added, and DNA is precipitated with 3 volumes of 100% ethanol. Electrophoretic analysis is then carried out using a DNA sequencing gel.

5. PSORALEN PHOTOBINDING TO MEASURE GLOBAL SUPERCOILING *IN VIVO*

5.1 Introduction

The binding of intercalating drugs such as Me$_3$-psoralen to supercoiled DNA is thermodynamically favoured, since binding unwinds DNA and relaxes negative supercoils. Thus, measurement of the rate of Me$_3$-psoralen photobinding to DNA can provide a reliable estimate of the level of unrestrained supercoiling (torsional tension) in both bacterial and eukaryotic chromosomes [4]. Upon irradiation with 360-nm light, psoralen forms monoadducts with bases as well as interstrand crosslinks. The binding of Me$_3$-psoralen is sufficiently weak that neither dark binding nor photobinding will displace nucleosomes or significantly perturb DNA topology [50].

The basic approach for detecting unrestrained supercoiling by this method is to measure the rate of photobinding to DNA under conditions in which (1) the DNA is expected to contain superhelical tension, and (2) the DNA is expected to be relaxed. The relative rate of Me$_3$-psoralen binding is nearly linearly proportional to the level of superhelical tension as shown by a standard curve generated under defined conditions *in vitro*. There are two ways to relax supercoils in *E. coli* cells. One is by treatment with drugs such as coumermycin [34], a specific inhibitor of DNA gyrase. The second is by introduction of nicks into DNA by treatment with ^{60}Co-irradiation [4]. Both approaches have been used to provide direct evidence that DNA in living *E. coli* cells contains unrestrained torsional tension [4,6]. This general approach has also shown that the linear chromosome of bacteriophage T4 in *E. coli* is under torsional tension [7] and that changes occur in the level of DNA supercoiling during sporulation in *Bacillus brevis* [51]. Since intracellular conditions cannot be duplicated *in vitro*, the estimate of *in vivo* superhelical density should be expressed in terms of the *in vitro* conditions used to generate the standard curve.

5.2   Me₃-psoralen photobinding to DNA in bacterial cells

5.2.1   Background

Photobinding in cells is essentially as described previously except that radioactive Me_3-psoralen is usually added near the saturation point (0.6 µg/mL). Following irradiation, DNA is purified [4]. The following procedure has been used to determine the specific activity of ³H-Me_3-psoralen-labelled DNA, which is then related to supercoiling by a standard curve prepared *in vitro*.

5.2.2   Buffers and Reagents

M9 buffer
    1 g NH_4Cl
    5.8 g Na_2HPO_4
    3.0 g KH_2PO_4
    dissolve in 1 L of distilled water

TEN buffer
    10 mM Tris-HCl pH 7.6
    50 mM NaCl
    1 mM Na_2EDTA

Tris/sucrose
    30% sucrose in 0.6 M Tris-HCl pH 8.1

Lysozyme
    10% lysozyme (in water)

EDTA
    32 mM Na_2EDTA

5.2.3   Procedure

1. Cells (4×10^9) are first washed with M9 buffer and then resuspended in 1.5 mL TEN buffer at 0–4°C.
2. Next, add 0.3 mL of ice cold Tris-sucrose, 0.2 mL of lysozyme, and 0.5 mL 32 mM EDTA. This suspension is incubated for 10 min at 0–4°C, followed by addition of sodium dodecyl sulphate to 0.5% to lyse the spheroplasts.
3. The volume of the lysate is adjusted to 3.5 mL with TEN, and the DNA is sheared by vortex mixing at high speed for 30 s.
4. The sample is incubated at 60°C for 10 min and extracted three

times with phenol and three times with chloroform-isoamyl alcohol (24:1).

5. Nucleic acids are precipitated by adjusting the solution to 0.3 M NaOAc, adding two volumes of 100% ethanol, and incubating at –20°C. They are then collected by centrifugation and redissolved in 0.3 mL TEN.

6. Pancreatic ribonuclease (10 μg) and T1 ribonuclease (5 U) are added, and the sample is incubated at least 2 h at 37°C to digest the RNA.

7. To separate the DNA and digested RNA, the samples are applied to a Bio-Gel A15 column (1 × 27 cm) equilibrated with TEN, and eluted (the DNA elutes with the void volume). Specific activities of the fractions are determined by measuring A_{260} and radioactivity.

5.3 Me$_3$-psoralen Photobinding to DNA in Eukaryotic Cells

5.3.1 Buffers and Reagents

PBS
    0.137 M NaCl
    2.7 mM KCl
    8.1 mM Na$_2$HPO$_4$
    1.5 mM NaH$_2$PO$_4$
    pH 7.5
NTE buffer
    10 mM Tris-HCl pH 7.6
    100 mM NaCl
    5 mM Na$_2$EDTA

5.3.2 Procedure

Adduct and crosslink formation is identical to that described for bacterial cells except that incubation with Me$_3$-psoralen and irradiation at 360 nm is carried out in PBS (without Ca^{2+} or Mg^{2+}) rather than in M9 buffer. Cells are then washed once and resuspended in PBS at about 5×10^6 cells/mL, and 1% (v/v) of a saturated Me$_3$-psoralen solution in ethanol is added [4]. For tissue culture cells attached to a plate, growth medium is removed, cells are washed with PBS, and 1–2 mL of PBS-containing Me$_3$-psoralen is added. These cells are then irradiated directly on an open plate.

The procedure for determining the specific activities of ^{3}H-Me$_3$-psoralen labelled DNA in eukaryotic cells is identical to that described above for bacteria except that after photobinding cells are resuspended in NTE buffer. They are then lysed by addition of 0.5% sodium dodecyl sulphate and 1 mg/mL proteinase K followed by incubation at 60°C for 60 min. Subsequent steps are identical to those used for analysis with bacterial cells.

6.   CONCLUDING REMARKS

Our ability to measure supercoiling has allowed us to examine the control of supercoiling, since linking deficit measurements reflect changes in the enzymes responsible for supercoiling. Two important concepts have emerged: (1) supercoiling is controlled; and (2) the overall level of supercoiling can change with growth environment. Additionally, we have learned that translocation processes can significantly alter local levels of supercoiling. This can even be seen inside living cells using torsionally tuned probes. We next expect to be able to experimentally perturb restrained supercoiling to better understand its basis.

Since DNA supercoiling is a controlled parameter that changes, we can begin to think about how it affects the activities of DNA. For example, it has long been known from *in vitro* studies that promoter activities are especially sensitive to changes in supercoiling. An attractive hypothesis is that promoter structure for specific genes has evolved to fit with the basal level of supercoiling experienced under a specific set of conditions. For example, a gene expressed at higher levels under aerobic than under anaerobic conditions may have evolved to be more active at lower levels of superhelical tension (supercoiling is lower under aerobic conditions [30,52]). The action of proteins that regulate transcription would be superimposed on the effects of superhelical tension. We are currently exploring the idea that this basal effect of supercoiling is exerted on promoters by influencing the orientation of their −10 and −35 sites of RNA polymerase binding.

So far, it has been difficult to make detailed statements about how supercoiling influences the activities of particular genes. In part, this is due to our inability to easily measure local supercoiling *in vivo* without perturbing either supercoiling or the structure of the gene of interest. Intercalating agents, such as Me$_3$-psoralen, will alter supercoiling when they bind to DNA if used at high concentration, and insertion of torsionally-tuned probes into genes may disrupt normal gene function. In addition, supercoiling is likely to affect many genes in addition to the one of interest. Thus, the direct effect of supercoiling may be on a regulatory gene rather than on the gene of interest. In spite of these ambiguities, the current methods for measuring supercoiling are making it possible to generate a substantial list of genes likely to be affected by supercoiling.

ACKNOWLEDGMENTS

We thank Marila Gennaro and Craig Benham for helpful comments on the manuscript. The authors' work is supported by grants from NIH (AI33337 and GM37677).

REFERENCES

[1] Laundon, C. and Griffith, J. Curved helix segments can uniquely orient the topology of supertwisted DNA. Cell, 52 (1988) 545–549.

[2] Wang, J.-Y. and Syvanen, M. DNA twist as a transcriptional sensor for environmental changes. Molecular Microbiol., 6 (1992) 1861–1866.

[3] Pettijohn, D.E. and Pfenninger, O. Supercoils in prokaryotic DNA restrained *in vivo*. Proc. Natl. Acad. Sci. USA, 77 (1980) 1331–1335.

[4] Sinden, R.R., Carlson, J. and Pettijohn, D.E. Torsional tension in the DNA double helix measured with trimethylypsoralen in living *E. coli* cells. Cell, 21 (1980) 773–783.

[5] Worcel, A. and Burgi, E. On the structure of the folded chromosome of *Escherichia coli*. J. Mol. Biol., 71 (1972) 127–147.

[6] Sinden, R.R. and Pettijohn, D.E. Chromosomes in living *Escherichia coli* cells are segregated into domains of supercoiling. Proc. Natl. Acad. Sci. USA, 78 (1981) 224–228.

[7] Sinden, R.R. and Pettijohn, D.E. Torsional tension in intracellular bacteriophage T4 DNA: Evidence that a linear DNA can be supercoiled *in vivo*. J. Mol. Biol., 162 (1982) 659–677.

[8] Gellert, M., O'Dea, M.H., Mizuuchi, K. and Nash, H. DNA gyrase: an enzyme that introduces superhelical turns into DNA. Proc. Natl. Acad. Sci. USA, 73 (1976) 3872–3876.

[9] Wang, J.C. Interaction between DNA and an *Escherichia coli* protein. J. Mol. Biol., 55 (1971) 523–533.

[10] Drlica, K. Control of bacterial DNA supercoiling. Molecular Microbiol., 6 (1992) 425–433.

[11] Pruss, G.J., Manes, S.H. and Drlica, K. Escherichia coli DNA topoisomerase I
      mutants: increased supercoiling is corrected by mutations near gyrase genes.
      Cell, 31 (1982) 35–42.
[12] DiNardo, S., Voelkel, K.A., Sternglanz, R., Reynolds, A.E. and Wright, A.
      Escherichia coli DNA topoisomerase I mutants have compensatory mutations
      in DNA gyrase genes. Cell, 31 (1982) 43–51.
[13] Amouyal, M. The remote control of transcription, DNA looping and DNA
      compaction. Biochimie, 73 (1991) 1261–1268.
[14] Drlica, K., Malik, M. and Rouviere-Yaniv, J. Intracellular supercoiling in bac-
      teria. In: Lilley, D.M.J. (ed.), *Nucleic Acids and Molecular Biology*, Springer-
      Verlag, Berlin, 1992, pp. 55–66.
[15] Rahmouni, A.R. and Wells, R.D. Stabilization of Z DNA *in vivo* by localized
      supercoiling. Science, 246 (1989) 358–363.
[16] Rahmouni, A.R. and Wells, R.D. Direct evidence for the effect of transcription
      on local DNA supercoiling *in vivo*. J. Mol. Biol., 223 (1992) 131–144.
[17] Cook, D., Armstrong, G. and Hearst, J. Induction of anaerobic gene expression
      in *Rhodobacter capsulatus* is not accompanied by a local change in chromoso-
      mal supercoiling as measured by a novel assay. J. Bacteriol., 171 (1989)
      4836–4843.
[18] Wells, R., Amirhaeri, S., Blaho, J., Collier, D., Dohrman, A., Griffin, J.,
      Hanvey, J., Hsieh, W., Jaworski, A., Larson, J., McLean, M., Rahmouni, A.,
      Rajagopalan, M., Shimizu, M., Wohlrab, F. and Zacharias, W. Biology and
      chemistry of Z DNA and triplexes. In: Drlica, K. and Riley, M. (eds.), *The
      Bacterial Chromosome*, American Society for Microbiology, Washington
      D.C., 1990, pp. 187–194.
[19] Thompson, R., Davies, J., Lin, G. and Mosig, G. Modulation of transcription
      by altered torsional stress, upstream silencers and DNA binding proteins. In:
      Drlica, K. and Riley, M. (eds.), *The Bacterial Chromosome*, American Society
      for Microbiology, Washington D.C., 1990, pp. 227–240.
[20] Depew, R. and Wang, J. Conformational fluctuations of DNA helix. Proc.
      Natl. Acad. Sci. USA, 72 (1975) 4275–4279.
[21] Pulleyblank, D., Shure, M., Tang, D. and Vinograd, J. Action of nicking-clos-
      ing enzyme on supercoiled and non-supercoiled closed circular DNA forma-
      tion of a Boltzmann distribution of topological isomers. Proc. Natl. Acad. Sci.
      USA, M2 (1975) 4280–4284.
[22] Keller, W. Determination of the number of superhelical turns in simian virus
      40 DNA by gel electrophoresis. Proc. Natl. Acad. Sci. USA, 72 (1975)
      4876–4880.
[23] Keller, W. and Wendel, I. Stepwise relaxation of supercoiled SV40 DNA. Cold
      Spring Harbor Symp. Quant. Biol., 39 (1974) 199–208.
[24] Pruss, G.J. DNA topoisomerase I mutants: Increased heterogeneity in linking
      number and other replicon-dependent changes in DNA supercoiling. J. Mol.
      Biol., 185 (1985) 51–63.
[25] Pruss, G.J. and Drlica, K. Topoisomerase I mutants: the gene on pBR322 that
      encodes resistance to tetracycline affects plasmid supercoiling. Proc. Natl.
      Acad. Sci. USA, 83 (1986) 8952–8956.
[26] Lockshon, D. and Morris, D.R. Positively supercoiled plasmid DNA is pro-
      duced by treatment of *Escherichia coli* with DNA gyrase inhibitors. Nucleic
      Acids Res., 11 (1983) 2999–3016.
[27] Liu, L. and Wang, J. Supercoiling of the DNA template during transcription.
      Proc. Natl. Acad. Sci. USA, 84 (1987) 7024–7027.

[28] Pruss, G.J. and Drlica, K. DNA supercoiling and prokaryotic transcription. Cell, 56 (1989) 521–523.

[29] Drlica, K., Franco, R. and Steck, T. Rifampicin and *rpo*B mutations can alter DNA supercoiling in *Escherichia coli*. J. Bacteriol., 170 (1988) 4983–4985.

[30] Hsieh, L., Burger, R.M. and Drlica, K. Bacterial DNA supercoiling and [ATP]/[ADP]: changes associated with a transition to anaerobic growth. J. Mol. Biol., 219 (1991) 443–450.

[31] Ish-Horowicz, D. and Burke, J. Rapid and efficient cosmid cloning. Nucleic Acids Res., 9 (1981) 2989–2998.

[32] Treisman, R. Transient accumulation of c-fos RNA following serum stimulation requires a conserved 5′ element and c-fos 3′ sequences. Cell, 42 (1985) 889–902.

[33] Sambrook, J., Fritsch, E. and Maniatis, T. Molecular Cloning, Cold Spring Harbor Press, Cold Spring Harbor, N.Y. (1989).

[34] Drlica, K. and Snyder, M. Superhelical *Escherichia coli* DNA: relaxation by coumermycin. J. Mol. Biol., 120 (1978) 145–154.

[35] Manes, S.H., Pruss, G.J. and Drlica, K. Inhibition of RNA synthesis by oxolinic acid is unrelated to average DNA supercoiling. J. Bacteriol., 155 (1983) 420–423.

[36] Steck, T.R., Pruss, G.J., Manes, S.H., Burg, L. and Drlica, K. DNA supercoiling in gyrase mutants. J. Bacteriol., 158 (1984) 397–403.

[37] Pruss, G.J., Franco, R., Chevalier, S., Manes, S.H. and Drlica, K. *Escherichia coli* topoisomerase I mutants: effects of inhibitors of DNA gyrase. J. Bacteriol., 168 (1986) 276–282.

[38] Cummings, D. Sedimentation and biological properties of T-phages of *Escherichia coli*. Virology, 23 (1964) 408–418.

[39] Hinton, D. and Bode, V. Ethidium binding affinity of circular lambda deoxyribonucleic acid determined fluorometrically. J. Biol. Chem., 250 (1975) 1060–1070.

[40] Hinton, D. and Bode, V. Purification of closed circular lambda deoxyribonucleic acid and its sedimentation properties as a function of sodium chloride concentration and ethidium binding. J. Biol. Chem., 250 (1975) 1071–1079.

[41] Zheng, G., Kochel, T., Hoepfner, R.W., Timmons, S.E. and Sinden, R.R. Torsionally tuned cruciform and Z-DNA probes for measuring unrestrained supercoiling at specific sites in DNA of living cells. J. Mol. Biol., 221 (1991) 107–120.

[42] Sinden, R.R. and Ussery, D.W. Analysis of DNA structure *in vivo* using Psoralen photobinding: Measurements of supercoiling, topological domains, and DNA-protein interactions. Methods Enzymol., 212 (1992) 319–335.

[43] Kochel, T.J. and Sinden, R.R. Analysis of trimethylpsoralen photoreactivity to Z-DNA provides a general in vivo assay for Z-DNA: Analysis of the hypersensitivity of (GT)n B-Z junctions. Biotechniques, 6 (1988) 532–543.

[44] Kochel, T.J. and Sinden, R.R. Hyperreactivity of B–Z junctions to 4,5′,8-trimethylpsoralen photobinding assayed by an exonuclease III/photoreversal mapping procedure. J. Mol. Biol., 205 (1989) 91–102.

[45] Maxam, A.M. and Gilbert, W. Sequencing end-labeled DNA with base-specific chemical cleavages. Methods Enzymol., 65 (1980) 499–559.

[46] Zagursky, R.J., Baumeister, K., Lomax, N. and Berman, M.L. Rapid and easy sequencing of double-stranded DNA and supercoiled DNA. Gene Anal. Tech., 2 (1985) 89–94.

[47] Cartwright, I.L. and Kelly, S.E. Probing the nature of chromosomal DNA-protein contacts by *in vivo* footprinting. BioFeature, 11 (1991) 188–203.

[48] Axelrod, J.D. and Majors, J. An improved method for photofootprinting yeast genes *in vivo* using Taq polymerase. Nucleic Acids Res., 17 (1989) 171–183.

[49] Hoepfner, R.W. and Sinden, R.R. Amplified primer extension assay for psoralen photoproducts provides a sensitive assay for a $(CG)_6TA(CG)_2(TG)_8$ Z-DNA torsionally tuned probe: preferential psoralen photobinding to one strand of a B–Z junction. Biochemistry 39 (1993) 7542–7548.

[50] Cimino, G.D., Isaacs, S.T. and Hearst, J.E. Psoralens as photoactive probes of nucleic acid structure and function: organic chemistry, photochemistry, and biochemistry. Annu. Rev. Biochem., 54 (1985) 1151–1193.

[51] Bogh, A. and Ristow, H. Tyrocidine-induced modulation of the DNA conformation in *Bacillus brevis*. Eur. J. Biochem., 170 (1987) 253–258.

[52] Dorman, C., Barr, G., NiBhriain, N. and Higgins, C. DNA supercoiling and the anaerobic growth phase regulation of *ton*B gene expression. J. Bacteriol., 170 (1988) 2816–2826.

Chapter 16

GEL RETARDATION ASSAYS AND DNA FOOTPRINTING

Ing Swie Goping, D. Alan Underhill and
Gordon C. Shore

OUTLINE

Methods in Gene Technology, Volume 2, pages 281–300
Copyright © 1994 JAI Press Ltd
All rights of reproduction in any form reserved.
ISBN: 1-55938-264-3

1.   INTRODUCTION

Defining the nature and outcome of protein–nucleic acid interactions within the promoters of protein-coding genes has been fundamental to our understanding of how initiation of transcription by RNA polymerase II is regulated in eukaryotes (reviewed in [1,2]). In this context, both gel retardation assays [3,4] and *in vitro* footprinting with DNase I [5] have been invaluable for characterizing the specific association of transcription factors with their cognate *cis*-elements. Gel retardation works under the simple premise that the mobility of a DNA fragment will be retarded in a low ionic strength polyacrylamide gel when bound to a protein. The success of DNase I footprinting derives from the observation that when protein binds to DNA, the sugar-phosphate backbone in this region of the DNA helix is no longer accessible to enzymatic cleavage by DNase I. Most routinely, DNase I footprinting is used to identify specific sequences within gene fragments that interact with nuclear proteins, and gel retardation then permits the detailed analysis of the protein-binding properties associated with specific *cis*-elements.

The DNase I footprinting analyses are usually carried out under conditions in which the concentration of the specific DNA target is limiting and the data obtained, therefore, is largely qualitative. In contrast, the gel retardation assay can be done with a large excess of probe, because free probe and protein-bound DNA are separated during electrophoresis. As a result, this assay can provide quantitative physical data related to DNA–protein interaction, notably association and dissociation rate constants [4]. Additionally, by varying the location of the recognition element relative to the ends of the DNA, changes in the mobility of the protein–DNA complex may occur as a consequence of conformational changes in the DNA-helix [6]. Furthermore, relative affinities for different *cis*-elements can be obtained by comparing the ability of various unlabelled DNA fragments to compete with the labelled probe for the same protein binding sites [7].

The gel retardation assay can be exploited to determine the protein binding specificity of a particular *cis*-element, by competing with oligonucleotides of known protein binding specificity. In this respect, it is also possible to reveal the identity of the protein species at a particular *cis*-element by incorporating specific antibodies (directed against the DNA-binding protein) in the gel shift assay [8]. If this protein is part of the DNA complex, there are two possible outcomes: (i) the antibody impedes binding of the DNA fragment and the protein–DNA complex does not form; and/or (ii) the immunoglobulin binds to the protein–DNA complex and retards its mobility to an even greater extent (gel supershifts).

The interaction of proteins with DNA in both DNase I footprinting and gel retardation analyses is sensitive to reaction conditions, and variations in pH, NaCl, KCl, $MgCl_2$, temperature, non-ionic detergents and non-specific competitor DNA can all influence protein binding. Unfortunately, DNase I footprinting normally must accommodate the binding of several distinct transcriptional regulatory proteins and it is not possible to simultaneously optimize conditions for all. However, gel retardation of a single *cis*-element is amenable to such changes, and doing so may allow the resolution of multiple protein DNA complexes.

A factor in the success of these experiments is the source of nuclear protein. We routinely obtain high quality, transcriptionally active nuclear protein extracts from whole tissue using the protocol of Gorski *et al.* [9] with the modification described by Maire *et al.* [10]. Similarly, extracts from cultured cells can effectively be prepared [11,12]. In addition, experiments are routinely performed using heat denatured extracts to reduce the complexity of binding species and to identify heat stable binding activities. Various binding activities can also be resolved using protein extracts that have been fractionated by ion-exchange chromatography.

2.   PRINCIPLE

The gel retardation assay protocols that are described here, are an adaptation of the method of Gilman et al. [13]. In these experiments, the putative protein binding site is end-labelled with [32]P. Nuclear proteins are allowed to bind to the DNA and the reaction is then subjected to polyacrylamide gel electrophoresis. The migration of the DNA–protein complex is retarded in comparison to the migration of the free probe, and the resulting pattern is seen on an autoradiogram of the dried gel. To test for the specificity of the protein–DNA interactions, excess unlabelled DNA is added as cold competitor.

For the in vitro footprinting assays, which are an adaptation of the method of Lichtsteiner [14], the DNA of interest is end-labelled on one end with [32]P. Proteins are allowed to complex to the DNA after which the DNA is digested at accessible sites by DNase I, at a frequency of approximately once per molecule. Therefore, the position of cleavage for each fragment of DNA relative to the ^{32}P labelled end, can be measured by migration and analysis of each radiolabelled species on a denaturing polyacrylamide gel. In the autoradiogram of the resulting gel, footprinted regions will appear as blank areas in the digestion pattern of the DNA in the presence of proteins, in comparison to the digestion pattern generated by naked DNA. The exact boundaries of protein-binding sites can be determined by comparison to a positional standard prepared by chemical cleavage of the probe adjacent to deoxyguanosine and deoxyadenosine residues [15].

3.   APPROACH

3.1   Choosing the Method for [32]P Labelling of Probes

The probe for mobility shift analysis can be obtained by purification from restriction enzyme-digested recombinant plasmid. More routinely, however, synthetic oligonucleotides corresponding to both strands of the region of interest are generated and are allowed to anneal together to form a double-stranded complex. If the fragment of DNA has 5 overhanging sequences, the probe can be generated by radiolabelling with Klenow in the presence of α-[32]P dNTP. Otherwise, the enzyme T4 polynucleotide kinase is used in the presence of γ-^{32}P ATP to incorporate ^{32}P at the ends of DNA fragments containing 5′-hydroxyl groups. Klenow end-labelled probes are generally more desirable if the extracts contain endogenous phosphatase activity.

Although the probe for gel retardation assays can be labelled on both ends, a footprinting probe must only have ^{32}P incorporated on one end of each probe molecule. For footprinting probes, the region of interest is usually obtained as a fragment from a restriction digest. Therefore, one end of the DNA is labelled with ^{32}P using either Klenow in the presence of α-^{32}P dNTP or kinase in the presence of γ-^{32}P ATP, and then the DNA is digested again with another restriction enzyme to create a probe that is labelled on one end. Alternatively, the polymerase chain reaction (PCR) can be used to create a footprinting probe by amplifying DNA in the presence of one primer that has been labelled with ^{32}P and one primer that is unlabelled.

3.2 Preparation of Buffers and Solutions

10 × Klenow buffer
  0.5 M Tris-Cl (pH 7.6)
  0.1 M MgCl$_2$

10 × kinase buffer
  0.5 M Tris-Cl (pH 9.5)
  0.1 M MgCl$_2$
  1 mM DTT

10 × band shift buffer
  250 mM HEPES (pH 7.6)
  50 mM MgCl$_2$
  340 mM KCl
  50% (v/v) glycerol

10 × footprinting buffer
  250 mM HEPES (pH 7.6)
  50 mM MgCl$_2$
  340 mM KCl

10 × vanadate buffer
  20 mM Na$_3$VO$_4$
  10 mM KF

Band shift acrylamide stock
   29 g acrylamide
   1 g *bis* acrylamide
   Make up to 100 mL with water. Filter and store in the dark at 4°C
Sequencing acrylamide stock
   38 g acrylamide
   2 g *bis* acrylamide
   Make up to 100 mL with water. Filter and store in the dark at 4°C
5 × glycine buffer
   30.28 g Tris
   142.64 g Glycine
   1.86 g EDTA
   Make up to 1 L with water
10 × CIP buffer
   10 mM $ZnCl_2$
   10 mM $MgCl_2$
   100 mM Tris-Cl (pH 8.3)
10 × PCR buffer
   500 mM KCl
   100 mM Tris-Cl (pH 8.4)
   6 mM $MgCl_2$
10 × TBE
   54 g Tris
   27.5 g boric acid
   3.72 g EDTA
   Make up to 500 mL with water. The pH will be approximately 8.3
Low salt wash
   50 mM Tris-Cl (pH 8.0)
   0.15 M NaCl
   10 mM EDTA (pH 8.0)
High salt wash
   50 mM Tris-Cl (pH 8.0)
   1 M NaCl
   10 mM EDTA (pH 8.0)
DB
   25 mM HEPES (pH 7.6)
   0.1 mM EDTA
   40 mM KCl

10% (v/v) glycerol

1 mM DTT

DNase I dilution buffer

25 mM CaCl$_2$

10 mM HEPES (pH 7.6)

Footprint stop solution

100 mM Tris-Cl (pH 7.5)

10 mM EDTA

100 mM NaCl

0.1% (w/v) SDS

Just before use, add yeast tRNA and proteinase K (self-digested) to a final concentration of 100 μg/mL and 1 mg/mL, respectively

10% APS (w/v)

0.1 g ammonium persulphate

Make up to 1 mL with H$_2$O

Formamide loading dye

80% (v/v) deionized formamide

50 mM Tris-borate (pH 8.3)

1 mM EDTA

0.1% (w/v) xylene cyanol

0.1% (w/v) bromophenol blue

3.3 Preparation of Polyacrylamide Gels

Band shift polyacrylamide gel: For a gel of approximately 40 mL in volume, combine:

8 mL band shift acrylamide stock

8 mL 5 × glycine buffer

375 μL 10% (w/v) APS

37.5 μL TEMED

23.5 mL H$_2$O

Cast the gel and let it polymerize for at least one hour.

Native TBE polyacrylamide gel: These gels are useful for visualization of DNA fragments that are too small to be clearly resolved on an agarose gel, and they are therefore used to gel purify footprinting probes. For a gel of approximately 40 mL in volume, combine:

4 mL 10 × TBE
7.5 mL band shift acrylamide stock
25.6 mL H$_2$O
375 µL 10% (w/v) APS
37.5 µL TEMED
Cast the gel and let it polymerize.

Urea sequencing gel: For a gel of approximately 120 mL in volume, combine:
57.6 g urea
48 mL H$_2$O
12 mL 10 × TBE
24 mL sequencing acrylamide stock
Dissolve the urea by swirling the gel solution at 37°C. Then add 110 µL of 20% APS and 110 µL of TEMED. Cast the gel and let it polymerize for at least 1 h.

4.   GEL RETARDATION ASSAYS

4.1   Labelling of the Probe

4.1.1   Klenow End-filling (Adaptation of Sambrook et al., [16])

1. Combine:
    1 µL (2 pmol/µL) double stranded oligonucleotide
    1.5 µL 10 × Klenow buffer
    1 µL 2 mM each dNTP minus the dNTP that is labelled with ^{32}P
    10 µL α-^{32}P dNTP (3000 Ci/mmol)
    1 µL water
    0.5 µL Klenow (5 U/µL)
    Incubate at 22°C for 20 min.
2. Add 1 µL of 2 mM cold dNTP corresponding to the α-^{32}P dNTP and incubate at 22°C for an additional 10 min.
3. Extract the DNA once with phenol/chloroform (phenol : chloroform : isoamyl alcohol at 50 : 49 : 1).
4. Purify the radiolabelled probe from the free nucleotides by filtration through a Sephadex G-50 column (Pharmacia Nick Column; 17-0855). The resulting concentration of the probe will be approximately 5 fmoL/µl at 5000 cpm/fmol.

4.1.2 Kinase End-labelling

1. Combine:
   1 μL (2 pmol/μL) double stranded oligonucleotide
   1 μL 10 × kinase buffer
   2 μL γ-^{32}P ATP (3000 Ci/mmol)
   5.5 μL water
   0.5 μL T4 polynucleotide kinase (8U/μL)
   Incubate at 37°C for 10 min.
2. Extract the DNA once with phenol/chloroform.
3. Purify the ^{32}P radiolabelled probe from the free nucleotides by filtration through a Sephadex G-50 column. The resulting concentration of probe will be approximately 5 fmol/μL.

4.2 Binding Reactions

This protocol is an adaptation of Gilman *et al.* [13]. Before preparing the binding reactions, pre-run the band shift polyacrylamide gel for approximately 1 h at 10 V cm^{-1} at 4°C. Assemble all the reagents on ice for the binding reaction.

1. Combine:
   1 μL probe (approx. 25 000 cpm/μL)
   n μL extract (1–20 μg)
   1 μL 10 × band shift buffer
   1 μL 10 × vanadate buffer
   1 μL 2 mg/ml poly (dI-dC) poly (dI-dC) (Pharmacia; 27-7880)
   Make up to 10 μL with water
   Incubate at 22°C for 15 min.
2. Stop the gel that has been pre-running. Directly load the samples into the wells. On outside lanes that do not contain band shift reactions, load some bromophenol blue that has been diluted in 5% (v/v) glycerol so that the progress of the gel can be visualized. Continue the electrophoresis at 10 V cm^{-1} at 4°C.
3. Stop the migration of the gel when the bromophenol blue is positioned approximately two-thirds from the top of the gel. Take apart the plates and place dry blotting paper on the surface of the gel. By peeling back the paper, the gel will be transferred from the glass plates to the paper, with a minimum of distortion. Dry the gel in a gel dryer for approximately 1 h at 80°C, under vacuum.

4. After the gel is dried, expose to film. For a probe that has been labelled to 5 000 cpm/µL, a 2 h exposure to Kodak X-Omat AR film should be sufficient. (see Figure 1)

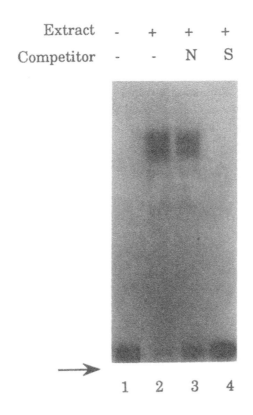

Figure 1. Representative autoradiograph of a gel retardation assay. The oligonucleotide has been end-labelled with T4 polynucleotide kinase in the presence of ^{32}P γ-ATP. The arrow indicates the position of the unbound probe. Lane 1 contains 5 fmol of probe alone, while lanes 2–4 contain 5 fmol of probe each in addition to 5 µg of rat liver nuclear extract. The binding reactions in lanes 3 and 4 were conducted in the presence of 500 fmol of non-specific oligonucleotide (N), and specific self oligonucleotide (S), respectively. Complexes that are formed specifically between the probe and the nuclear proteins are competed by the addition of excess unlabelled oligonucleotide (lane 4). The gel was exposed for 3 h at –70°C.

5.   DNA FOOTPRINTING

5.1   Labelling of the Probe

5.1.1   Klenow End-filling

1. Linearize 10 μg of plasmid (approx. 2 pmol of a 6.5 kb plasmid) with a restriction enzyme, generating a 5′-overhang. This is the site that will become the ^{32}P labelled end of the footprinting probe. Incorporate α-^{32}P dNTP at the ends of the linearized DNA using the previously described method for Klenow-end filling of oligonucleotides for generation of mobility shift probes. (see Section 4.1.1)

2. Digest the linearised DNA with a second restriction enzyme that cuts the original DNA into asymmetric fragments, thereby defining the unlabelled end of the footprinting probe. Purify the probe of interest by preparative gel electrophoresis. (see Section 5.2)

   If the second restriction site is very close to the first, the smaller second fragment will not be efficiently recovered by ethanol precipitation. Therefore either of these two procedures will result in a relatively purified probe preparation.

   These procedures will generate a specific DNA fragment labelled only on one end with ^{32}P, that will be used as a footprinting probe.

3. Alternatively, create a fragment of DNA with one terminus that can be labelled with a specific α-^{32}P dNTP in the presence of Klenow, such that the other end of the duplex molecule remains unlabelled. Purify the probe from the unincorporated nucleotides by ethanol precipitation or gel filtration.

5.1.2   Kinase End-filling

For fragments containing blunt ends or 3′-overhangs, dephosphorylate the DNA [16], then label with T4 polynucleotide kinase in the presence of γ-^{32}P ATP.

1 Combine:

   10 μL digested DNA (approx. 10 μg of DNA)
   2.5 μL of 10 × CIP buffer
   11.5 μL of water
   1 μL calf intestinal alkaline phosphatase (CIP), (1 U/μL)
   Incubate at 55°C for 30 min.

2. Add a further 1 µL of CIP and incubate for an additional 30 min.

3. Stop the reaction by adding EDTA to a final concentration of 5 mM and inactivate the CIP by heating to 75°C for 10 min, followed by extraction with phenol/chloroform and precipitation of the DNA.

4. Incorporate γ-³²P at the 5′ end of each DNA strand, following the protocol for kinase-labelling of oligonucleotides for generation of mobility shift probes. (see Section 4.1.2)

5. Digest the DNA with a second restriction enzyme, to generate a specific DNA fragment labelled at only one end with ³²P. Separate the fragments as described in Section 5.2.

5.1.3 Generating a probe with PCR

Design 2 primers (A and B) that will produce the DNA fragment of interest upon PCR amplification, in the presence of the appropriate recombinant plasmid.

1. End-label 9 pmol of primer A using the previously described polynucleotide kinase protocol (see Section 4.1.2, except in this case, the oligonucleotide will be single-stranded and you will need to increase the amount of γ-³²P ATP to 4 µL).

2. PCR amplify the region encompassing the footprinting probe by combining:

   3 µL γ-³²P end-labelled primer A (approx. 1 pmol/µL)

   1 µL unlabelled primer B (3 pmol/µL)

   5 µL 10 × PCR buffer

   1 µL dNTP stock (0.5 mM each dNTP)

   1 µL plasmid (10 ng/µL)

   39 µL of H_2O

   0.5 µL Taq polymerase (5 U/µL)

   Incubate in a thermal cycler, with annealing temperatures that correspond to the melting point of the primers. For example, for the amplification of a 400 bp fragment by 2 primers (18-mer each) with roughly 50% G+C content, 17 cycles of 94°C for 1 min, 55°C for 1 min and 72°C for 1 min, followed by a final step of 72°C for 10 min is usually successful.

5.2 Purification of the Probe

If the probe was generated by end-labelling of a restriction enzyme product, the DNA of interest must by purified from the ^{32}P labelled vector sequences and from unincorporated nucleotides. If the probe was generated by PCR reaction, the probe must be purified from the ^{32}P labelled primers and from unincorporated nucleotides. Various methods are available for the purification of DNA from acrylamide gel slices. The electrophoresis into DEAE membranes and subsequent elution is described and is an adaptation of the manufacturer's instructions.

1. Prepare both a native 6% TBE polyacrylamide gel (see Section 3.3), and a 0.8% (w/v) agarose gel in $1 \times TBE$. The wells of the agarose gel should be slightly larger than the wells of the polyacrylamide gel.

2. Add loading dye to the sample and migrate the probe on the polyacrylamide gel until the bromophenol blue dye is at the mid-point of the gel.

3. Carefully wrap the gel in plastic wrap and tape it to a piece of paper. Dot the paper at three places with radioactive ink and expose the gel to X-ray film for approximately 5 min.

4. After developing the film, align the film with the gel and the radioactive markers and cut out the section of the gel that matches the position of the electrophoresed probe.

5. With blunt forceps, slide the excised gel slice into the wide wells of the agarose gel. Cut a piece of DEAE membrane (NA 45, Schleicher and Schuell; 23410) to the dimensions of the gel slice, dampen with $1 \times TBE$ and carefully place the membrane against the acrylamide gel slice so that the DNA will migrate towards the membrane in an electrophoretic field.

6. After approximately 30 min, determine if most of the radioactive probe has left the gel slice by checking the radioactivity of the gel slice with a hand-held monitor. If the level of radioactivity has significantly decreased, remove the DEAE membrane and wash it briefly in low salt wash.

7. Place the membrane in a 1.5 mL microfuge tube containing 400 µL of high salt wash. Incubate at 68°C for 30 min.

8. Remove the solution to a fresh tube and incubate the DEAE membrane with 100 µL of high salt wash for an additional

15 min. Combine the elution solutions and precipitate. There will be approximately 1 pmol of labelled probe. Dilute an aliquot of the probe to obtain a final concentration of 1 fmol/μL at approximately 5000 cpm/fmol.

5.3   DNase I Titrations

In this example of a set of footprinting reactions [14], the DNase I digestion pattern of the probe in the absence of nuclear extract will be compared to the digestion pattern in the presence of 5 μg, 10 μg, and 40 μg of nuclear extract proteins. The ability of DNase I to digest DNA is altered in the presence of extract, so in the initial experiment the optimal amount of DNase I needed for each reaction will be determined.

1.   Set up 4 sets of 4 microfuge tubes each, labelled 1 to 16. Each set of tubes will contain 0, 5 10 and 40 μg of extract that will each be exposed to 5, 10, 25 and 75 ng of DNase I.

2.   Prepare a premix solution that contains:
     1.7 μL 10 mg/mL poly (dI-dC) poly (dI-dC)
     34 μL 10 × footprint buffer
     34 μL 10 × vanadate buffer
     17 μL probe
     117.3 μL H$_2$O
     Aliquot 12 μL of premix in all 16 tubes.

3.   Add no extract to tubes 1 to 4, 5 μg of extract to the next four tubes (5 to 8), 10 μg of extract to tubes 9 to 12, and 40 μg of extract to tubes 13 to 16. The initial extract concentration should be at least 5 mg/mL so that the total volume of extract added does not exceed 8μL. Add DNase/RNase-free BSA (Pharmacia; 27-8914) so that each tube contains 40 μg of total protein. Add DB, so that the final volume in each tube is 20 μL. Incubate at room temperature for 15 min then place tubes on ice.

4.   Prepare DNase I (BRL; 8047SA) dilutions of 5, 10, 25, and 75 ng/μL in DNase I dilution buffer. Let them stand on ice for 5 min.

5.   Add 1 μL of 5 ng/μL DNase I to tubes 1, 5, 9 and 13.
     Add 1 μL of 10 ng/μL DNase I to tubes 2, 6, 10 and 14.
     Add 1 μL of 25 ng/μL DNase I to tubes 3, 7, 11 and 15.

Add 1 μL of 75 ng/μL DNase I to tubes 4, 8, 12 and 16.

Incubate each reaction on ice for 3 min exactly, then add 80 μL footprint stop solution. Stagger the addition of DNase to each tube, so that the probe in each sample is digested for exactly 3 min.

6. Incubate the samples in the stop solution at 37°C for 30 min. Extract with phenol/chloroform. Precipitate with ethanol and wash with 70% ethanol.

7. Resuspend in 5 μL of formamide loading dye. Incubate for 2 min at 90°C and migrate on a 8% polyacrylamide 7 M urea sequencing gel (see Section 3.3) that had been pre-run at approximately 2000 V until a temperature of 50°C was reached. Continue electrophoresis at 2000 V at 50°C.

8. After the bromophenol blue has reached the bottom of the gel, stop the electrophoresis. Dis-assemble the apparatus and fix the gel in 10% acetic acid : 10% methanol for approximately 15 min. Carefully transfer the gel to a piece of blotting paper, cover with plastic wrap and dry on a gel dryer for 1 h at 80°C under vacuum. Expose the gel to Kodak X-Omat AR film with intensifying screens.

9. After an overnight exposure, develop the film. Determine the DNase I dilution for each extract concentration that resulted in bands of approximately equal intensity, appearing throughout the length of the gel. Also, assess the appearance of the undigested probe and choose DNase I dilutions that will result in equal probe intensity at the top of the gel. See Figure 2 for an example of the amount of DNase I used in footprinting reactions containing variable amounts of added extract. Proceed to Section 5.4.

5.4 Footprinting Reactions

1. Each reaction will contain:
   0.1 μL of 10 mg/mL poly (dI-dC) poly (dI-dC)
   2 μL of 10 × footprint buffer
   2 μL of 10 × vanadate buffer
   1 μL probe
   6.9 μL of H_2O
   Make a premix solution and aliquot 12 μL of the premix into each tube.

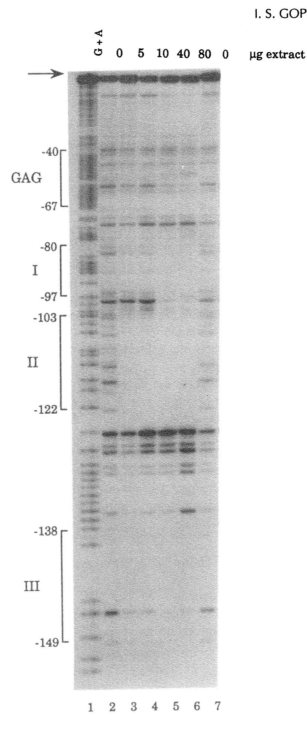

2. Add appropriate amounts of extract to each tube. Adjust the amount of final protein with BSA and complete the volume in each tube to 20 µL with DB. Allow the binding reactions to occur for 15 min at 22°C, then place samples on ice.

3. Prepare DNase I dilutions in DNase I dilution buffer and let stand on ice for 5 min.

4. Add the amount of DNase I that had been empirically determined in Section 5.3, to each tube and incubate the reactions on ice for exactly 3 min before addition of footprint stop buffer.

5. Incubate in the stop solution at 37°C for 30 min and then extract with phenol/chloroform. Precipitate with ethanol and wash with 70% ethanol. Resuspend in 5 µL of formamide loading dye. Heat for 2 min at 90°C and migrate (2000 V at 50°C) on a 8% polyacrylamide 7 M urea sequencing gel (see Section 3.3).

6. After the bromophenol blue has reached the bottom of the gel, stop the electrophoresis. Fix the gel in 10% acetic acid : 10% methanol and dry on a gel dryer for 1 h at 80°C under vacuum. Expose the gel to Kodak X-Omat AR film with intensifying screens overnight. If the reactions and gel were successful, re-expose the gel for a few days without any intensifying screens in order to get sharper bands. (see Figure 2)

Figure 2.   Representative autoradiograph of a footprinting gel. The probe which is approximately 170 bp in length has been end-labelled with T4 polynucleotide kinase in the presence of ^{32}Pγ-ATP. The arrow indicates the position of the undigested probe. Lane 1 contains Maxam and Gilbert G+A sequencing reactions of the same probe (see [16]). The amounts of rat liver nuclear extract added to each reaction are indicated along the top of the gel. Reactions that contained no extract (lanes 2 and 7) or 5 µg of extract (lane 3) were digested with 5 ng of DNase I. The sample containing 10 µg of extract was digested with 10 ng of DNase I (lane 4). The sample containing 40 µg of nuclear extract was digested with 22 ng of DNase I (lane 5). The sample containing 80 µg of nuclear extract was digested with 75 ng of DNase I (lane 6). Footprinted regions are indicated to the left of the gel, in brackets. The gel was exposed for 5 d at –70°C.

6. TROUBLE SHOOTING

6.1   Mobility Shift Probe is Smeary

If the oligonucleotides are not purified, the ^{32}P-labelled probe will not migrate as a single band but will appear as a series of bands above and below the major radiolabelled species.

Gel purify the unlabelled oligonucleotide by migration on a preparative 8% 7 M urea sequencing gel. (see [16]). Other causes of poor band clarity may result from the gel heating up during electrophoresis, decreasing the resolution of the bands. Make sure that the gel is run at 4°C.

6.2   Loss of Radioactivity of Mobility Shift Probes

The nuclear extract may contain endogenous phosphatase activity. If the probe has been labelled with T4 polynucleotide kinase, be sure to add vanadate to the binding reactions. Otherwise, end-label the oligonucleotide with Klenow.

6.3   Footprinting Gels Look Blurry

If the footprinting gels have been exposed in the presence of fast intensifying screens, such as Quanta III (DuPont; 323101), the bands will be blurry. Expose the gel without any screens and use Kodak X-Omat RP film. The exposure time will need to be increased about 10 fold. However, this results in much sharper bands.

6.4   Footprinting Gels are Blank

As in Section 6.2, the nuclear extract may contain endogenous phosphatase activity.

If the probe has been labelled with T4 polynucleotide kinase, be sure to add vanadate to the binding reactions. Otherwise, end-label the oligonucleotide with Klenow. There may also be nuclease activity in the BSA used to normalize the amount of total protein in each reaction. Be sure to use DNase/RNase-free BSA such as Pharmacia (27-8914).

Alternatively, the probe may be over-digested. Repeat the DNase I titrations (Section 5.3) using lower concentrations of DNase I. Always use RNase-free DNase such as BRL (8047SA) RQ1 DNase I.

6.5 Footprinting Gels are Blank at the Top of the Gel

The probe has been over-digested. Repeat the reactions with lower concentrations of DNase I.

6.6 Footprinting Gels are Blank at the Bottom of the Gel

The probe has been under-digested. Repeat the reactions with higher concentrations of DNase I.

6.7 Footprinting Bands become Narrow at the Bottom of the Gel

There is some residual SDS in the sample.
Always wash the DNA from the final precipitation step, with 70% ethanol, prior to heating and loading on the sequencing gel.

6.8 Level of Probe Intensity at Top of Gel is Uneven

The probe has not been digested to the same relative amounts in each sample.
Adjust the amount of DNase I accordingly, for each reaction.

ACKNOWLEDGMENTS

We would like to thank Dr. C. Mueller and Dr. M. Lagacé for invaluable help with these techniques. G.S. is a recipient of operating grants from the Medical Research Council of Canada and the National Cancer Institute of Canada. I.S.G. is a recipient of a Steve Fonyo Research Studentship from the National Cancer Institute of Canada. D.A.U. is a recipient of a fellowship from the Cancer Research Society of Canada.

REFERENCES

[1] Maniatis, T., Goodbourn, S. and Fischer, J. Regulation of inducible and tissue-specific gene expression. Science, 236 (1987) 1237–1245.

[2]   Mitchell, J.M. and Tjian, R. Transcriptional regulation in mammalian cells by sequence-specific DNA binding proteins. Science, 245 (1989) 371–378.

[3]   Garner, M.M. and Revzin, A. A gel electrophoresis method for quantifying the binding of proteins to specific DNA regions. Nucl. Acids Res., 9 (1981) 3047–3060.

[4]   Fried, M.G. and Crothers, D.M. Equilibria and kinetics of lac repressor-operator interactions by polyacrylamide gel electrophoresis. Nucl. Acids Res., 9 (1981) 6505–6525.

[5]   Galas, D. and Schmitz, A. DNase footprinting: a simple method for the detection of protein-DNA binding specificity. Nucl. Acids Res., 5 (1978) 3157–3170.

[6]   Zinkel, S.S. and Crothers, D.M. DNA bend direction by phase sensitive detection. Nature, 328 (1987) 178–181.

[7]   Chodosh, L.A., Carthew, R.W. and Sharp, P.A. A single polypeptide possesses the binding and activities of the adenovirus major late transcription factor. Mol. Cell. Biol., 6 (1986) 4723–4733.

[8]   Kristie, T.M. and Roizman, B. α4, the major regulatory protein of herpes simplex virus type 1, is stably and specifically associated with promoter-regulatory domains of a genes and/or selected viral genes. Proc. Natl. Acad. Sci. U.S.A., 83 (1986) 3218–3222.

[9]   Gorski, K., Carneiro, M. and Schibler, U. Tissue-specific *in vitro* transcription from the mouse albumin promoter. Cell, 47 (1986) 767–776.

[10] Maire, P., Wuarin, J. and Schibler, U. The role of *cis*-acting elements in tissue-specific albumin gene transcription. Science, 244 (1989) 343–346.

[11] Ohlsson, H. and Edlund, T. Sequence-specific interactions of nuclear factors with the insulin gene enhancer. Cell, 45 (1986) 35–44.

[12] Dignam, J.D., Lebovitz, R.M. and Roeder, R.G. Accurate transcription initiation by RNA polymerase II in a soluble extract from isolated mammalian nuclei. Nucl. Acids Res., 11 (1983) 1475–1489.

[13] Gilman, M.Z., Wilson, R.N. and Weinberg, R.A. Multiple protein-binding sites in the 5′-flanking region regulate c-fos expression. Mol. Cell. Biol., 6 (1986) 4305–4316.

[14] Lichtsteiner, S., Wuarin, J. and Schibler, U. The interplay of DNA-binding proteins on the promoter of the mouse albumin gene. Cell, 51 (1987) 963–973.

[15] Maxam, A.M. and Gilbert, W. Sequencing end-labelled DNA with base-specific chemical cleavages. Methods Enzymol., 65 (1980) 499–560.

[16] Sambrook, J., Fritsch, E.F. and Maniatis, T. *Molecular Cloning. A Laboratory Manual.* Cold Spring Harbor Laboratory Press, Cold Spring Harbor, N.Y., 1989.

Chapter 17

SOUTHWESTERN BLOT

Giorgio Corte and Maria T. Corsetti

OUTLINE

Methods in Gene Technology, Volume 2, pages 301–317
Copyright © 1994 JAI Press Ltd
All rights of reproduction in any form reserved.
ISBN: 1-55938-264-3

1. INTRODUCTION

The gel retardation, or band shift assay, described in Chapter 16 is a very good method to detect DNA-binding proteins, to assess their specificity for a particular sequence and even to measure their binding affinity. However, by this technique nothing can be learned about the characteristics of the molecules involved. The extent of retardation is not proportional to the mass of the protein nor does the number of retarded bands reflect the number of different DNA-binding proteins actually present. The Southwestern blot, on the other hand, can establish not only the molecular weight of the binding protein, but also the number of different proteins involved and, to some extent, provide information on their characteristics. However, this latter technique is not as sensitive and cannot be used to measure binding affinity. The Southwestern blot is then not an alternative to the gel retardation assay, but rather is complementary to it.

The method is not entirely new, being a combination of SDS-PAGE with already existing methods of detecting DNA binding proteins on nitrocellulose filters; this combination provides a screening technique that facilitates the cloning of genes encoding sequence-specific DNA-binding proteins [1,2].

Electrophoresis in SDS, unlike the electrophoretic run in non denaturing buffers used in the gel retardation assay, will cause the dissociation of multimeric proteins; therefore the method will be useless if the DNA-binding factor under study requires two different polypeptide chains to form the sequence specific binding site.

Although many eukaryotic proteins that bind DNA in a sequence-specific manner can withstand the denaturation-renaturation process, the resilience of each individual DNA-binding factor to this treatment should be assessed in advance. This step is indeed crucial. Not only will a poor renaturation reduce the sensitivity of the technique to unacceptably low levels, but it could also affect the sequence specificity of the protein leading to meaningless results. Even if the protein can be renatured with success, the yield will not be one hundred percent and the individual molecules will be tightly packed in a narrow band. The signal obtained will therefore be always lower than that observed in a gel retardation experiment with the same amount of protein.

The use of catenated probes consisting of a high density of protein-binding sites labelled at high specific activity may be helpful but steric hindrance will still be a problem. It is then important to try different methods of denaturation-renaturation to find the procedure that gives the best results for the protein studied [2–8]. The guanidine hydrochloride procedure has worked for many DNA-binding proteins and may be of general use [2,6]. However one must keep in mind that each protein has its unique properties and that even highly homologous proteins do not necessarily behave in the same way.

With these limitations, the technique is still useful when different DNA-binding proteins with the same or similar specificity are coexpressed in the same cell, as in the case of homeobox proteins, as it allows the unequivocal identification of their number and identity. Similarly, in the presence of different forms (due, for instance, to posttranslational modifications including glycosylation and phosphorylation), this technique enables the form(s) that have DNA-binding activity to be identified. In our experience the method has also been very useful for following DNA-binding proteins during purification. Once the molecular weight of the protein has been determined, it is possible to unmistakably recognise it even in the presence of other DNA-binding proteins. Unless specific antibodies are available, a Southwestern blot is the only way to make sure that transfected cells are indeed producing the protein in its correct form.

2. PRINCIPLE

The first step, the preparation of the sample, should not be underestimated. Enough DNA-binding protein must be loaded on to the gel or nothing will be seen, no matter how carefully and skillfully the other steps of the procedure are carried out. Ideally, it is desirable to be able to load the total cell lysate directly dissolved in a denaturing agent. However, DNA-binding proteins are usually not abundant and quite often some purification and/or concentration step will be necessary, as the method is not extremely sensitive and there is a limit to the amount of material that can be loaded on a gel.

Great care must then be taken and all available inhibitors of proteolysis should be used to preserve the integrity of the protein. For the same reason, whenever possible, the nuclear extract should be prepared by dissolving the nuclei directly in loading buffer or in a denaturing agent. Many DNA-binding proteins are not satisfactorily solubilized in the usual buffers used, for instance, for the gel retardation assay. On the other hand, since the Southwestern blot requires the complete denaturation of the protein anyway , there is no sense in risking the loss of the protein for the sake of using the same buffer. Our advice is to use the 8 M urea buffer whenever possible. In our experience the best results are obtained when the sample (total cell lysate, nuclear extract, column fraction or purified protein) is dissolved in this buffer prior to the addition of the SDS loading buffer.

The following step, SDS-PAGE, is performed to separate the proteins according to their apparent molecular weight and does not require any special procedure. The separated proteins are then transferred to a nitrocellulose filter as for a normal Western blot. The proteins must now be renatured while bound to the filter. They are then first fully denatured to destroy any non-specific folding generated during the previous steps and then slowly renatured to give the polypeptide chain the best chance to refold properly forming the correct disulphide bonds. Again, this step should be carried out with great care, because, if not enough protein refolds in its native tertiary structure, the amount of probe bound in the last step will be undetectable.

The last step, the DNA-binding step, does not involve any special

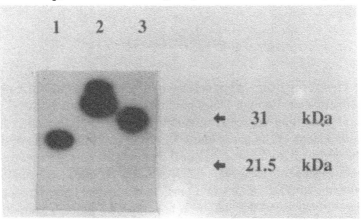

Figure 1. Autoradiograph of a Southwestern blot of three human recombinant homeobox proteins produced in the Baculovirus system. (1) HOX 2C; (2) HOX 4B; (3) HOX3C. The three proteins show specific binding to BS2-18, an oligonucleotide whose sequence is derived from the promoter of Antennapedia [13].

problems and is carried out under conditions similar to the incubation of the gel retardation assay. These are actually quite different from those of a Western or a Southern blot. An example of the results that can be achieved is shown in Figure 1.

3 APPROACH

3.1 Preparation of the Sample

The sample to be analysed by Southwestern blot can be obtained from a variety of sources, according to the particular DNA-binding protein under study and the purpose of the investigator, for instance identification of the molecule, determination of its characteristics, or purification. The most common sources are: total cell lysates, nuclear extracts of eukaryotic cells, bacterial cells, samples taken after each purification step (especially ion-exchange or affinity chromatography fractions), immunoprecipitated material, products of '*in vitro*' translation systems, and recombinant proteins. Depending on the starting material, different strategies can be used to prepare the sample for Southwestern blots.

3.1.1 *Tissues or Cell Pellets*

If it is impossible to disrupt the cells and isolate the nuclei, as in the case of frozen material or when it is known that the protein is very labile, the best way to proceed is to lyse the whole sample by boiling after addition of an equal volume of 2 ×SDS-PAGE loading buffer. Alternatively, nearly complete solubilization can be achieved resuspending the pellet in 8 M urea plus 1% β-mercaptoethanol (see Section 3.2). A short sonication can be necessary to obtain satisfactory solubilization in this latter buffer. Long sonication times may damage the protein of interest and conditions will have to be adjusted for each tissue or cell line. As an indication, in our hands 30–60 s at 40C at an amplitude of 14–16 microns in a MSE Soniprep 150 are sufficient for most cells.

Guanidine salts are good solubilizing agents but should be avoided because they form insoluble precipitates with SDS. Unless the DNA-binding protein is relatively abundant, as in the case of recombinant proteins produced in bacteria or eukaryotic cells, its concentration in a sample prepared in this simple way will not be high enough to allow its detection in Southwestern blots.

3.1.2   Nuclear Extracts of Eukaryotic Cells

The isolation of cell nuclei is a simple and rapid technique that results in a five-fold purification and concentration of nuclear proteins. The concentration of most DNA-binding proteins in a nuclear extract is sufficient for the Southwestern blot. However, for the less abundant ones, further subfractionation will be necessary to achieve the required concentration. Purification of the nuclei can be performed in two ways:

(1) The cells are resuspended in hypotonic buffer (see Section 3.2) at a maximum concentration of 5 x 108 cells/mL, transferred to a Dounce homogenizer and homogenized with the tight fitting pestle until all the cells are broken and the nuclei are separated from the cytoplasm. The number of the strokes varies for different types of cells. The lysis should be controlled every few strokes under the microscope mixing 5 μL of lysate with 5 μL of Trypan blue [9]. The isolated nuclei are then pelleted by centrifugation at 100–200g for 5 min and the supernatant is discarded.

(2) As an alternative procedure, the cells are washed twice with PBS, resuspended in PBS/0.5% NP40 at a maximum concentration of 108 cells/mL and incubated for 5 min in ice. The nuclei are then pelleted by centrifugation at 100–200g for 5 as in the other method. However, since the nuclear membrane is dissolved by NP40, the naked nuclei tend to clump together and are often impossible to resuspend in aqueous buffers. For further subfractionation, 2 mL of cell lysate are layered on a 6 mL cushion of 0.25 M sucrose/10 mM MgCl2 in tubes (Falcon 2051) and centrifuged at 100 g for 5 min. Besides affording a better separation between nuclei and cellular debris, this method yields a nuclear pellet which can be readily resuspended in buffer C.

The nuclei can be directly loaded on the gel after lysis by boiling in SDS-PAGE loading buffer after resuspension in 8 M urea plus 1% ß-mercaptoethanol or they can be extracted as follows:

1.   The nuclei are resuspended in buffer C, 5×108 nuclei/mL, transferred to a Dounce homogenizer and homogenized with the tight fitting pestle (15 strokes). Alternatively, if a sonicator is available, the nuclei can be sonicated until complete

disruption.

2. The nuclear homogenate is kept in ice for 30 min and then centrifuged at 1000g for 10–30 min to remove insoluble material.

3. The resulting supernatant can be loaded on the gel after boiling in SDS PAGE loading buffer. Alternatively, an enriched fraction can be obtained by successive elutions of the nuclear pellet with increasing salt concentrations [for an example see ref. 10].

3.1.3   Bacterial Cells

1. 1 mL of culture is centrifuged at 10 000 g for 1–2 min, the supernatant is discarded, and the cell pellet is resuspended in 1 mL of ice cold water and centrifuged at 10000g for 2 min.

2. After discarding the supernatant, the cells are resuspended by vortexing in 100 μl of water. An equal volume of 2 × SDS gel loading buffer is added to the suspension and mixed with a yellow pipette tip.

3. The cell lysate is boiled for 5–10 min.

4. Chromosomal DNA must be sheared by sonication, otherwise the solution is too viscous; the power and the length of the sonication depend on the type of the sonicator. Sonicate until the fluidity is sufficient to load the sample with a micropipette [11].

3.1.4.   Chromatographic Fractions

As pointed out in the introduction, the Southwestern blot can be used to follow a DNA-binding protein during the purification steps. Ion-exchange chromatography is the most widely used method to purify nuclear proteins. If the fractions are recovered in a buffer that is well tolerated in SDS-PAGE, 50 μL of 4 × SDS gel loading buffer are added to 150 μL of each fraction, the samples are boiled for 5 min and directly loaded on the gel.

If the fractions are in a buffer that is not well tolerated, that is containing calcium or very high concentrations of monovalent ions, the sample must be dialysed against a suitable buffer.

3.1.5   Immunoprecipitated Material

Immunoprecipitates can be directly resuspended in SDS PAGE loading buffer, boiled and loaded.

3.1.6   Products of in vitro *Translation Systems*

Usually the amount of protein synthesized in vitro is too low to allow detection in Southwestern blots. However, if the translation system has worked particularly well, the reaction mixture can be directly loaded after boiling in SDS PAGE loading buffer. Most often fractionation and concentration of the reaction mixture will be necessary.

3.2   Preparation of Solutions and Gel

4 x SDS-PAGE gel loading buffer
   2 mL SDS 20%
   2 mL glycerol
   5 mL solution 3 (see below)
   100 µL Bromophenol blue 1%
   1 mL 2-mercaptoethanol

Solution 1
   29.2 g acrylamide
   0.8 bisacrylamide
   Make up to 100 mL with distilled water and filter the solution through a 0.45 µm membrane.

Solution 2
   1 M Tris HCl pH 8.8
   12.1 g Tris
   1.15 mL HCl 37%
   Make up to 100 mL with distilled water and filter the solution through a 0.45 µm membrane.

Solution 3
   0.25 M Tris HCl pH 6.8
   3 g Tris
   2 mL HCl 37%
   Make up to 100 mL with distilled water and filter the solution through a 0.45 .µm membrane.

Tris-glycine 10×
   144 g glycine
   30 g Tris
   Make up to 1 L with distilled water.

Running buffer
 100 mL Tris-glycine 10×
 10 mL SDS 10%
 Make up to 1 L with distilled water.
Transfer buffer
 100 mL Tris-glycine 10×
 200 mL methanol
 Make up to 1 L with distilled water.
APS 10%
 1 g ammonium persulphate
 Make up to 10 mL with distilled water. It can be stored at 4°C for 1
 week
Hypotonic buffer
 10 mM Hepes pH 7.9
 10 mM $MgCl_2$
 10 mM KCl
 0.5 mM DTT
 5 μg/mL Aprotinin (see below)
 0.5 mM PMSF (see below)
Buffer C
 20 mM Hepes pH 7.9
 25% v/v glycerol
 1 mM $MgCl_2$
 0.42 M NaCl
 0.2 mM EDTA
 0.5 mM DTT
 5 μg/mL Aprotinin (see below)
 0.5 mM PMSF (see below)
Aprotinin (protease inhibitor)
 10 mg Aprotinin
 Dissolve in 1 mL of PBS. Store at -20°C avoiding repeated freezing
 and thawing.
PMSF (phenylmethylsulphonyl fluoride; serine protease inhibitor)
 10 mg PMSF
 Dissolve in 1 mL of isopropanol.
8M Urea/1% mercaptoethanol
 Add 0.5 mL of 1% β-mercaptoethanol to 50 mL of 8 M urea.
 Urea solutions must be made with enzyme grade urea and
 discarded after two weeks.

10 ×Ponceau S
  2 g Ponceau S
  30 g trichloracetic acid, or 30 mL of 100% trichloracetic acid solution
  30 g sulphosalicylic acid
  Make up to 100 mL with distilled water.
1 × Ponceau S (working solution)
  10 mL of 10 × Ponceau S
  90 mL of water
10 × PBS (phosphate buffered saline)
  8 g NaCl
  0.2 g KCl
  1.44 g Na_2HPO_4
  0.24 g KH_2PO_4
  Dissolve in 80 mL of distilled water.
  Adjust the pH to 7.2. Make up to 100 mL with distilled water.
1 × PBS (working solution)
  10 mL 10 × PBS
  90 mL of distilled water
PBS/0.5%NP40/10 mM $MgCl_2$
  10 mL 10 × PBS
  50 µl NP40
  100 µl of 1 M $MgCl_2$
  Make up to 100 mL with distilled water.
Trypan blue
  0.25 g Trypan blue
  10 mL 10 × PBS
  Make up to 100 mL with distilled water. Either filter-sterilize and handle aseptically, or add sodium azide to 0.02%
  CAUTION: care is needed in handling sodium azide.
BB (binding buffer)
  25 mM Hepes pH 7.9
  25 mM NaCl
  0.5 mM DTT
  1 mM $MgCl_2$
BB/6M Guanidine
  28.7 g guanidine hydrochloride
  Make up to 45 mL with distilled water
  Add the appropriate amounts of stock solutions (Hepes, $MgCl_2$, DTT, NaCl) as for 50 mL of BB. Make up to 50 mL with distilled water.

BB/1% milk

    Add 1 g of dried skimmed milk (Carnation or similar, 1.3% fat) to 100 mL of BB.

BB/0.25% milk

    Add 0.25 g of dried skimmed milk (Carnation or similar, 1.3% fat) to 100 mL of BB.

3.2.1 Preparation of the gel

The electrophoresis is carried out in slab gels in a discontinuous Tris buffer system [12]. The system consists of two gels: the lower (separating) gel and the upper (stacking) gel.

Several apparatuses for vertical slab gel electrophoresis are commercially available. Glass plates, spacers and combs must be cleaned with detergent and rinsed well. After complete polymerization of the lower gel, the upper gel is cast on top of it.

The gel volume varies slightly according to the different apparatus and should be calculated to allow 1–2 cm of upper gel below the bottom of the wells. The easiest way is to prepare an excess of solution, mark on the glass plate the desired height of the lower gel and pour the solution up to the mark.

The acrylamide concentration in the separating gel has to be chosen in order to achieve optimal separation in the molecular weight range expected for the DNA-binding protein under use. A 6% gel separates well between 200 and 50 kDa, while a 10% gel is suitable for molecular weights between 100 and 20 kDa. Nearly all proteins will fall in this range. If the protein is small, a 12% gel will extend the range down to about 10 kDa, while also increasing the separation below 30 kDa. Higher acrylamide concentrations, or even gradient gels, can be used, but the gel becomes rapidly brittle at high concentrations and special care will have to be taken in handling and drying.

The concentration of the upper gel, being only a stacking gel, is always the same, regardless of the concentration of the lower. The recipes are given for 10 mL of gel solution.

Lower gel:
For 6% gel mix:
    2 mL solution 1
    3.75 mL solution 2
    4 mL water

For 10% gel mix:
   3.3 mL solution 1
   3.75 mL solution 2
   2.7 mL water
For 12% gel mix:
   4 mL solution 1
   3.75 mL solution 2
   2 mL water
Degas the solution in a vacuum flask and add:
   0.1 mL SDS 10%
   5 μl TEMED
   0.1 mL APS 10% and pour between the gel plates.

The polymerizing solution is immediately overlaid with water-saturated 2-butanol to exclude air. Being lighter, 2-butanol neatly floats on top of the solution which will then polymerize with a perfectly straight edge. The gel is allowed to polymerize for between 30 min and 1 h.

Upper gel: When the lower gel is polymerized, the butanol is removed by tilting the gel apparatus, or by aspirating with a syringe.
Upper gel mix:
   1.5 mL solution 1
   5 mL solution 3
   3.3 mL water
Degas the solution under vacuum, and add:
   0.1 mL SDS 10%
   5 μl TEMED
   0.1 mL APS 10%
and pour on the lower gel up to the brim of the glass plates. The comb is then immediately inserted taking care not to trap air bubbles that would result in cavities in the teeth.

3.3   Additional Items

Dounce homogenizer with B pestle
Ultrasonic disintegrator

4.   METHOD

1.   Remove all the liquid left in the wells formed in the stacking

gel, load the samples with a micropipette and fill up the wells with running buffer. It is better to load prestained molecular weight markers in the lateral lanes to control the separation in the gel and the transfer to the filter.

2. Assemble the electrophoretic cell taking care not to disturb the loaded samples. Fill the lower and upper buffer reservoirs with running buffer.

3. Run at 4 W until the Bromophenol Blue tracking dye has reached the bottom of the gel.

4. Cut four sheets of Whatman 3MM paper and one sheet of nitrocellulose of the size of the gel; soak the nitrocellulose in H2O.

5. Disassemble the electrophoresis apparatus, recover the gel and rinse in transfer buffer.

6. Dip the nitrocellulose membrane, the filter papers and the support pads of transfer apparatus in transfer buffer.

7. Assemble the transfer sandwich in the following order:
   support pad
   two filter papers
   gel
   nitrocellulose
   two filter papers
   support pad

8. Remove air bubbles and make sure the sandwich is tightly assembled; place the complete sandwich in the transfer apparatus.

9. Transfer at 300 mA for 2–3 h with cooling or at 80 mA overnight.

10. After transfer, place the nitrocellulose filter in a glass dish or in a tray of appropriate size. The prestained molecular weight markers must appear completely transferred . Stain with 1 × Ponceau S solution for 10 s, destain with distilled water until the background is clear and make a Photocopy as a permanent record.

11. Wash out the remaining Ponceau S from the filter with distilled water. All the following steps must be carried out at 4°C and with chilled buffers.

12. Add just enough BB/6M Guanidine to cover the filter and incubate for 5 min with gentle shaking. Use a tray of the same size of the filter to reduce the volume of the buffer.

13. Decant the solution in a graduated cylinder, add an equal volume of BB (without guanidine!!) and replace in the tray with the filter; incubate with the diluted solution (now containing 3M guanidine) for 5 min.
14. Repeat step 13 five times.
15. Remove the solution and wash the filter with BB for 5 min with gentle agitation.
16. Remove the washing buffer and replace with BB/1% milk; incubate for 30 min.
17. Transfer the filter to a plastic bag, add BB/1% milk with 106 cpm/mL of radiolabelled probe. The probe is usually labelled by Klenow filling [11 and other chapters of this volume) or, if it is a synthetic oligonucleotide, by phosphorylation with T4 polynucleotide kinase (see Chapter 11 and other chapters in this volume).
18. Incubate for 2 h with gentle shaking.
19. Wash for 10 min with BB/0.25% milk.
20. Repeat step A two times.
21. Dry the filter briefly on Whatman 3MM paper, then wrap in a transparent plastic film and expose to an X-ray film at -70°C with an intensifying screen.

5. GENERAL POINTS

5.1 Radioactivity

Working with radioactive material requires appropriate precautions e.g. wearing disposable gloves and using suitable screen. Rules for handling and disposing of radioactive products differ depending on the country and institution. Check the local rules.

5.2 Storage

Although the best results are obtained with fresh samples, nuclear extracts can be stored at −80°C in buffer C. After boiling in SDS-PAGE loading buffer they can be stored at −20°C. After transfer, the filter can be stored wet at 4°C in binding buffer for a few days.

5.3 Electrophoretic Run

At 4 W, a gel 1.5 mm thick, 15 cm long and 13 cm wide (the size of most commercially available apparatuses) will run in 4–5 h. The intensity of the current drops during the run and unless the power pack has a constant power mode, one has to adjust the voltage accordingly every 20–30 min. The gel can be warm but must never become hot. If the gel apparatus has a cooling system, the power can be raised to 10 W, considerably shortening the running time. If it is more convenient, run the gel overnight at 40 V without cooling and raise the voltage to 150 V the next morning to complete the run.

6. TROUBLE SHOOTING

6.1 No Signal

(a) The amount of material is too low. Check that no proteolysis has occurred. If possible, try some kind of purification. A 3 mm thick gel will double the amount of protein that can be loaded on the gel without affecting transfer.

(b) The denaturation-renaturation step has not worked. Use only molecular biology grade guanidine-HCl. Try different conditions (e.g. salts and DTT concentration, pH).

(c) No binding. Was the probe prepared correctly? Were the correct wash temperature and duration employed?

6.2 High Background

Check that the filters did not dry at any step.

Check that the DNA probe does not contain radiolabelled single strand DNA.

Repeat the washing step.

If the background is normal but several bands are visible, this may be due to non specific binding which can be abolished by adding to the probe a 50–100 fold excess of non radioactive competitor DNA (usually poly(dI-dC)(dI-dC).

ACKNOWLEDGMENTS

This work was partially supported by grants from MURST and CNR Progetto Finalizzato Biotecnologie (BTBS)

REFERENCES

[1]   Singh H., LeBowitz J.H., Baldwin, A.S. and Sharp P.A. Molecular cloning of an enhancer binding protein: Isolation by screening expression library with a recognition site DNA. Cell, 52 (1988) 415–423

[2]   Vinson C.R., LaMarco K.L., Johnson P.F., Landschulz W.H., McKnight S.L. In situ detection of sequence-specific DNA binding activity specified by a recombinant bacteriophage. Genes Dev., 2 (1988) 801–806

[3]   Venkateswara Rao M., Rangarajan P.N. and Padmanaban G. Dexamethasone negatively regulates phenobarbitone-activated transcription but synergistically enhances cytoplasmic levels of cytocrhrome P-450b/e messenger RNA. J. Biol. Chem., 265 (1990) 5617–5622

[4]   Burnstein K.L., Jewell C.M. and Cidlowski J.A. Human glucocorticoid receptor cDNA contains sequences sufficient for receptor down-regulation. J. Biol. Chem., 265 (1990) 7284–7291

[5]   Knepel W., Jepeal L. and Habener J.F. A pancreatic islet cell-specific enhancer-like element in the glucagon gene contains two domains binding distinct cellular proteins. J. Biol. Chem., 265 (1990) 8725–8735

[6]   Herrera V.L.M. and Ruiz-Opazo N. Regulation of -Tropomyosin and N5 genes by a shared enhancer. J. Biol. Chem., 265 (1990) 9555–9562

[7]   Gunther C.V., Nye J.A., Bryner R.S. and Graves B.J. Sequence-specific DNA binding of the proto-oncoprotein ets-1 defines a transcriptional activator sequence within the long terminal repeat of the Moloney murine sarcoma virus. Genes Dev., 4 (1990) 667–679

[8]   Harrison M.J., Lawton M.A., Lamb C.J. and Dixon R.A. Characterization of a nuclear protein that binds to three elements within the silencer region of a bean chalcone synthase gene promoter. Proc. Natl. Acad. Sci. USA, 88 (1991) 2515–2519

[9]   Dignam J.D., Lebovitz R.M. and Roeder R.G. Accurate transcription initiation by RNA polymerase II in a soluble extract from isolated mammalian nuclei. Nucleic Acid Res., 11(1983) 1475–1489

[10] Odenwald W.F.,Garben J., Arnheiter H., Tournier-Lasserve E. and Lazzarini R.A. The Hox-1.3 homeo box protein is a sequence-specific DNA-binding phosphoprotein. Genes Dev., 3 (1989) 158–172

[11] Sambrook J., Fritsch E.F. and Maniatis T. Molecular Cloning. A Laboratory Manual. 2nd edn., Cold Spring Harbor Laboratory Press, Cold Spring Harbor, N.Y. 1989.

[12] LaemmLi U.K. Cleavage of structural proteins during the assembly of the head of bacteriophage T4. Nature, 227 (1970) 680–685

[13] Corsetti M.T., Briata P., Sanseverino L., Daga A., Airoldi I., Simeone A., Palmisano G., Angelini C., Boncinelli E. and Corte G. Differential DNA binding properties of three human homeodomain proteins. Nucleic Acid Res., 20 (1992) 4465–4472

Chapter 18

GENOMIC FINGERPRINTING USING THE PCR-RANDOM AMPLIFIED POLYMORPHIC DNA TECHNIQUE

S. V. Primal S. Silva and James D. Procunier

OUTLINE

Methods in Gene Technology, Volume 2, pages 319–336
Copyright © 1994 JAI Press Ltd
All rights of reproduction in any form reserved.
ISBN: 1-55938-264-3

1. INTRODUCTION

The polymerase chain reaction (PCR) uses enzymatic synthesis to exponentially amplify a specific DNA sequence and has enabled molecular biologists to examine a particular sequence of a chromosome [1, 2]. Since its development, PCR has revolutionized standard molecular biological techniques in the areas of genetic mapping [3], DNA fingerprinting [4, 13–14], forensic science [5], infectious disease diagnostics [6], and prenatal diagnosis of inherited disorders [7].

The ingredients for performing the PCR reaction include a source of DNA, heat stable DNA polymerase, various buffers, single nucleotides for DNA synthesis and primers 10 to 20 base pairs in length. The choice of primers is probably the most critical step since each one has a different Tm of annealing depending upon its base sequence. The primers are orientated with their 3´ends pointing towards each other and anneal to opposite DNA strands. Thus, the amplified segment is between the two primer sequences defined by the 5´ends of the primers and is generally 2000 base pairs or less in length.

The PCR method is based on repeating cycles, each composed of three basic steps:

(i) denaturing of DNA above 90°C;
(ii) lowering the reaction temperature to allow annealing of primers to opposite DNA strands; and
(iii) primer extension (72°C) whereby the complementary DNA strand is synthesized by Taq DNA polymerase.

The primers are physically incorporated into the PCR product. During the first cycle, the <long> product is formed, its length dependent on how long the polymerase is functional. However, on subsequent cycles, the DNA polymerase will terminate at the previous initiation point (the primer site) giving the <short> product. The <short> product then acts as a template resulting in its exponential increase in amount with each cycle. Thus a single molecule of template DNA can be amplified by several million-fold. The PCR products may be analysed directly on agarose or polyacrylamide gels and are stained with ethidium bromide yielding distinct band(s). In addition, the band(s) can be further characterized by either Southern blotting [8] and hybridization with a specific probe or by running restriction enzyme digestion products on gels [9].

The occurrence of natural DNA sequence polymorphisms has been detected by restriction fragment length polymorphisms (RFLP) analysis [10]. Restriction endonuclease digestion of total genomic DNA is followed by hybridization with a labelled probe in order to reveal differently sized hybridizing fragments. RFLPs result from either base pair changes within a restriction enzyme recognition site or deletions, insertions and chromosomal rearrangements within a restriction fragment. However, RFLP analysis is technically complex and time consuming. Thus, the feasibility of using RFLPs in high throughput diagnostic tests or crop breeding programs is questionable.

A specific class of molecular markers that detect polymorphisms are random amplified polymorphic DNAs (RAPDs) generated via the PCR technique [11,12]. Instead of using a pair of known sequence oligonucleotide primers to amplify a specific target sequence, a single 10-mer of an arbitrary sequence is used. This short primer binds to many different loci and amplifies on average 2–10 products as seen in bacteria [13], protozoa [14], fungi [15], plants [16,17], and humans [11]. An unlimited number of random sequence primers which contain at least 50% Gs and Cs and lack internal inverted repeats are available. Polymorphisms result from changes in either the primer binding site sequence (no amplification) or size differences in the amplified product (insertions, deletions). The products are easily separated by electrophoresis on standard agarose gels and visualized by ultraviolet (UV) illumination of ethidium bromide stained gels.

A low frequency of RFLP or RAPD markers may be observed when comparing species strains, races or certain plant cultivars [18]. One reason for the low frequency, aside from the narrow genetic base, is that both of these markers result from base changes within a target size of less than 10 bases. Base changes within the remaining portion of the amplified fragment cannot be detected using standard gel electrophoresis conditions. By utilizing denaturing gradient gel electrophoresis (DGGE), detection of a single base change between amplified products is possible. Differential melting properties between fragments alter the migration of DNA homoduplexes during electrophoresis [19,20]. Fragments of identical size can be distinguished and theoretically, up to 100% of the base changes within a fragment can be resolved under proper conditions. At a particular denaturant concentration, a DNA fragment becomes partially melted and undergoes a decrease in electrophoretic mobility. The region of the fragment that melts first is the early melting domain (generally, T_m = 71°C) and a later melting domain (Tm = 80°C) is usually present in most fragments. RAPD amplified fragments can be analyzed under DGGE conditions to give a superior method for detecting DNA polymorphisms [21].

The PCR-based DNA markers have numerous advantages over conventional RFLP, isoenzyme [22] or monoclonal antibody markers [23]. The procedure is technically simple, quick to perform and requires only small amounts (nanograms) of DNA. There is virtually an unlimited supply of random primers for polymorphism testing. The tests results are not affected by environmental factors or modifier genes found in different genetic backgrounds (epistasis). The markers found have been located in both single copy and repeated DNA, thus more of the genome is analyzed. The PCR procedure can be done at an early stage of development and is non-destructive (DNA analysed from a few cells). Thus, PCR-based markers are well suited to high throughput applications such as diagnostic testing or use in plant breeding programs.

2. MATERIALS

Lysozyme (Sigma)
   1 mg mL⁻¹ in water
TBSE
   100 mM Tris-Cl (pH 8.0)
   150 mM NaCl
   2 mM EDTA
SDS
   Sodium dodecyl sulphate (BioRad)
CTAB
   hexadecyltrimethyl-ammonium bromide (Sigma)
   prepare a 10% solution
Chloropane
   equal volumes of water saturated phenol (Fisher) and chloroform
   1% isoamyl alcohol
   0.01% hydroxy quinoline (Mathesan Coleman and Bell) adjust to
   pH 8.0 and store in a dark bottle

Proteinase-K (Boehringer Mannheim)
  10 mg mL⁻¹ in:
    100 mM Tris-HCl (pH 7.5)
    10 mM EDTA
  self-digested for 1 h at 50°C
TNE
  100 mM Tris-Cl (pH 8.0)
  50 mM EDTA
  1.4 M NaCl
RNase A (Sigma Type I-A, R4875)
  10 mg mL⁻¹ in water.
  Place in boiling water for 20 min
TKO 100 minifluorometer: (Hoefer Scientific Instruments, San Francisco, Calif)
Bisbenzimide
  Hoechst 33258 trihydrochloride (Sigma) (10 mg mL⁻¹ in water).
DNA assay solution
  10 mM Tris-Cl (pH 7.4)
  10 mM EDTA
  0.20 M NaCl
  bisbenzimide (diluted 1:10⁴)
  Filter the solution
InstaGene: purification matrix (BioRad)
IsoQuick: nucleic acid extraction kit (MicroProbe, Garden Grove, Calif)
Metal xanthates: potassium ethyl xanthogenate (Fluka)
Primers: (Biotechnology Laboratory, University of British Columbia, Vancouver, BC, Canada; Operon Technologies, Alameda, Calif.)
NAP-5 columns: (Pharmacia LKB)
10×PCR buffer, magnesium free
  500 mM KCl
  100 mM Tris-HCl (pH 9.0 at 25°C)
  1% Triton-X 100
Taq DNA polymerase from *Thermus aquaticus* strain YT1
dNTP solution
  100 mM solution of each dATP, dCTP, dGTP and dTTP neutralized to pH 7.0 (Boehringer Mannheim)

Formamide
   de-ionized by stirring gently with Dowex AG50-X-8 mixed bed
   resin for 30 min. Collect the formamide by filtering through
   Whatman 1 MM paper (BioRad)
Wax beads: Fisher Tissue Prep 2, T555, nominal cost
Temed: N, N, N', N'-tetramethylethyldiamine (BioRad)
pGEM markers: DNA fragments from 36 bp to 2.645 kb (Promega)
Ethidium bromide
   10 mg mL^{-1} in water, stored in a dark bottle
Stop buffer
   50 mM Tris-HCl (pH 8.0)
   50 mM EDTA
   1% SDS
   0.25% bromophenol blue
   0.25% xylene cyanol
   50% glycerol

3. METHODS

3.1 DNA Isolation

The isolation of high molecular weight DNA becomes a major
concern when performing RFLP analysis. However, smaller DNA
fragments are quite adequate when using PCR techniques. The fresh
material (bacteria, cells, plant tissue) can be lyophilized to total
dryness and stored with a desiccant indefinitely at –20°C.
   One of the difficulties in isolating DNA from bacteria or plants is
rupturing the rigid cell wall. Bacterial cell walls can *often* be removed
by lysozyme. Bacteria are pelleted by centrifugation and washed once
in TBSE. Upon resuspension in TBSE, the cells are treated with 1 mg
mL^{-1} of lysozyme for 10 min on ice. (More refractory bacterial species
may require a combination of detergent, heat and mechanical action to
break the cell wall). Plant cell walls are broken by grinding in liquid
nitrogen in a mortar and pestle. About 0.25 g of dry material with a
pinch of washed sand is ground for 30 sec or until a fine powder
results.
   Cell lysis and DNA purification follows the conventional
proteinase-K/phenol/SDS procedure. One modification of the

procedure is the addition of CTAB [24]. This effectively removes contaminating polysaccharides which can interfere with the PCR reaction. To 250 mg of material, add 20 mL of prewarmed (65°C) TNE buffer, make the solution 1% CTAB and add proteinase K (50 µg mL^{-1}). After mixing, SDS is added to a final concentration of 1% and the solution incubated for 2 h at 65°C. An equal volume of chloropane is added and the solution shaken for 30 min. The phases are separated by centrifugation for 10 min in a clinical centrifuge (2000 r.p.m.). The DNA in the upper phase is precipitated by adding 0.6 volumes of isopropanol and wound out on a glass rod. This leaves most of the RNA behind. The wound DNA is washed with 70% ethanol (squirt bottle) and then dissolved in 1–2 mL of distilled, deionized water. Any traces of phenol can be removed by an ether extraction. If significant amounts of RNA are present, this may inhibit the PCR reaction [25]. Addition of RNase A (10 µg mL^{-1}) for 3 h at 37°C will eliminate this problem. It is not necessary to remove the byproducts of RNase digestion for the PCR reaction.

Estimates of DNA concentration can be obtained by A_{260} measurements [26] and quality assessed by electrophoresis on 0.7% agarose gels. A more precise determination of DNA concentration without interference from contaminating RNA or protein is achieved by fluorescence measurements in the presence of bisbenzimide using a TKO 100 mini-fluorometer (excitation, 365 nm; emission, 460 nm). It is important to obtain accurate measurements of DNA concentration since the ratio of primer to DNA is critical for the PCR reaction.

For applications requiring high throughput, various DNA extraction methods have been reported for rapid extraction and using small amounts of material [27]. An added advantage is that the methods do not require the use of toxic phenol. Commercial nucleic acid extraction kits are also available (IsoQuick, Microprobe Corp.; InstaGene, BioRad). A non-grinding procedure for cell walled material has recently been reported [28]. The method uses metal xanthates to solubilize the polysaccharides of the cell wall.

3.2 PCR Amplification

3.2.1 Standard PCR

The primers used in the PCR-RAPD technique are random

oligonucleotides (10-mers) that have at least 50% G + C and have been purified by NAP-5 drip columns. It is recommended that all PCR manipulations be conducted in a sterile laminar flow hood in order to prevent contamination. A separate set of pipettes with positive displacement are needed and used only for loading the reactants. All additions are done on ice.

For simple DNA (bacteria), the following reagents are added in order in a 25 µL reaction volume:

1. distilled, deionized water, 13.5 µL
2. 10×PCR buffer, 2.5 µL
3. 25 mM MgCl$_2$, 1.5 µL
4. 2.5 mM of each nucleotide, 2.0 µL
5. 0.20 µM primer, 1.5 µL
6. 10 ng µL^{-1} DNA, 3.0 µL
7. 5.0 U µL^{-1} Taq DNA polymerase, 0.50 µL

The reaction volume is mixed and overlaid with 50 µL of mineral oil.

For more complex DNA (plants), the ratio of primer to DNA is increased by about three times. To avoid crystallization, the MgCl$_2$ solution is heated at 65°C for 10 min, vortexed and stored at 4°C for one month. The batch method eliminates pipetting variability by combining all ingredients except the desired variable ones (DNA or primer) for many samples. Negative controls are prepared as described above except that no DNA is added. Formamide can minimize background, thus improving the specificity of the reaction [29].

3.2.2 Hot Start PCR

The 'hot start' technique prevents complete mixing of PCR reactants until the reaction temperature is high enough to melt the wax bead. This minimizes nonspecific annealing of primers to non-target DNA and reduces the formation of primer dimers [30]. Thus, increases in the overall amplification specificity, precision and yield are achieved. Mineral oil is not used as a vaporization barrier. A solid wax layer is formed over a subset of PCR reactants and the remaining reactants then are added above the wax. During the first thermal cycle, the denaturing temperature melts the wax layer allowing thermal convection to completely mix the reactants. A reasonable ratio of the upper aqueous volume to lower volume is 3 : 1 to ensure complete mixing. The polymerase is separate from the primer and

MgCl$_2$ in order to prevent primer dimer formation.

For a 50 μL reaction volume, the lower volume (14 μL) contains:

1.   10×PCR buffer, 1.0 μL
2.   25 mM MgCl$_2$, 3 μL
3.   2.5 mM of each nucleotide, 5.0 μL
4.   0.20 μM primer, 5 μL.

A wax bead is added, the tube heated to 65°C for 10 min and then cooled. To the upper volume (36 μL), the following are added:

1.   distilled deionized water, 29 μL
2.   10×PCR buffer, 4 μL
3.   10 ng μl^{-1} DNA, 3 μL
4.   5.0 U μl^{-1} Taq polymerase, 0.50 μL.

The tube is then placed in the thermal cycler.

3.2.3   Thermocycle Conditions

Following an initial denaturation of 2 min at 94°C, the PCR is carried out for 45 cycles in a Perkin-Elmer Cetus (Norwalk, Conn.) TC-1 Thermal Cycler. The cycle program (using a 10-mer primer) is for 1 min at 94°C, 1 min at 36°C, and 2 min at 72°C. The PCR is followed by a final primer extension-polymerization of 10 min at 72°C.

3.3   Analysis of Amplified Fragments

3.3.1   Agarose Gels

The amplified DNA fragments (Figure 1) are mixed with stop buffer (10 : 1, vol/vol, respectively) and electrophoresed on 1.4% agarose gels run in standard 1×TAE buffer for 4 h at a constant 80 V. The gel is stained with 0.50 μg mL^{-1} of ethidium bromide and photographed under ultraviolet illumination with a red filter and Polaroid (Cambridge, Mass.) type 57 positive film. Promega pGem DNA fragments are used as molecular markers.

3.3.2   Denaturing Gradient Gel Electrophoresis (DGGE)

It is necessary to run the gels at a high temperature near the melting temperature of most naturally occurring DNA fragments. A temperature range between 60°C and 90°C is used to establish a relatively steep denaturing gradient in the gel. The vertical gel is

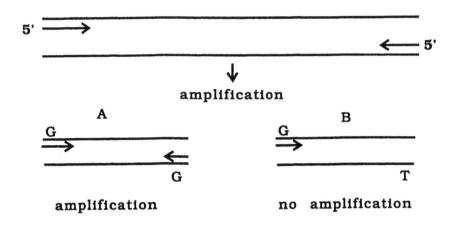

Figure 1.   RAPD marker polymorphism.

The RAPD primer is an oligonucleotide of a random sequence, 10 bases in length. Generally a single primer is used in the PCR reaction. The primer anneals to opposite denatured DNA strands of a target sequence. Annealing temperature is about 36°C and primer extension is from the 3´end (arrow). A natural DNA sequence polymorphism is shown between two different DNA samples (A and B). These may represent different species strains, races or plant cultivars. Amplification occurs for DNA sample A because the primer is complementary to the target sequence. For DNA sample B, the primer is not complementary due to the G to T transversion and no amplification is observed. Insertions and deletions at the primer binding site can also cause the same result. The agarose gel phenotype is the presence of a band (sample A) versus its absence (sample B).

submerged in a circulating water bath (Lauda) with a constant temperature of 60°C and the denaturants urea and formamide used to establish the gradient. Gels are run in a Bio-Rad Protean II apparatus using vertical 16 ×16 cm, 0.75 mm thick, 10% polyacrylamide. Fragment sizes between 0.10 and 1.0 kb are separated on this gel. A linear gradient of 20-60% denaturant (100% = 7 M urea, 40% formamide, v/v) is used with the highest concentration at the anode and the lowest at the cathode (Figure 2). A peristaltic pump is necessary to continually replace the buffer in the upper chamber from the lower chamber. The gel apparatus is placed on a stirrer with the bar in the lower chamber.

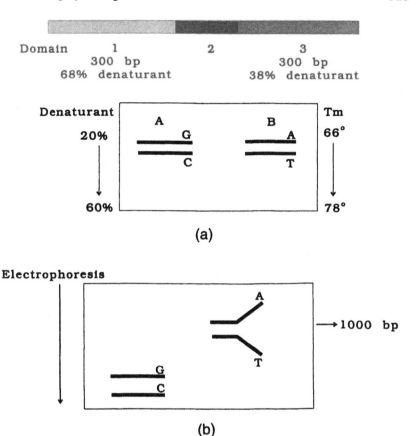

Figure 2. Melting behaviour of a RAPD fragment in a DGGE gel. (a) The top drawing represents a typical RAPD fragment at ambient temperature. The 1000 bp fragment melts in two domains with T_ms of 71°C (38% denaturant) and 80°C (68% denaturant). Melting domains are usually between 25 to 300 base pairs in length. The DNA fragments are electrophoresed in 10% polyacrylamide gels with a linear-increasing 20–60% denaturant gradient, equivalent to a temperature gradient of 66–78°C. On average, about one half of the length of the DNA fragment shows the lower temperature melting domain. By using the same single primer, two different DNA samples (A and B) are the source of RAPD fragments. The fragments differ by a single base in the early (lower) melting domain. (b) During electrophoresis for 1200 V h, the two fragments that differ by a single base are separated. The early melting domain of fragment B has melted to appear as a branched molecule. The partial melting of DNA fragments in DGGE causes mobility retardation which results in the DNA fragments to focus sharply, allowing a very fine resolution of bands.

The following procedure describes the reagents, denaturing gradient formation, and gel polymerization:

1. Acrylamide stock:
   40% acrylamide
   20 g acrylamide
   0.53 g bisacrylamide
   made up to 50 mL water
2. 20 ×TAE buffer (higher salt for DGGE)
   0.8 M Tris-HCl, pH 7.4
   0.4 M Na acetate
   0.02 M Na EDTA
3. Denaturant stock: 100% = 7M urea plus 40% formamide
   21.0 g urea
   20 mL deionized formamide
   made up to 50 mL with water
4. Denaturing gradient: use a 15–25 mL gradient maker

Chamber A (60% denaturant)		Chamber B (20% denaturant)
2.85 mL	acrylamide (40%)	2.85 mL
6.84 mL	denaturant (100%)	2.28 mL
0.57 mL	buffer (20x)	0.57 mL
1.14 mL	water	5.70 mL

The solutions are degassed on ice for 10 min. Temed (5.7 μl) is then added to each chamber. After addition of 20% ammonium persulphate (57 μL), the gradient is started using a long needle to reach the bottom of the gel in order to prevent mixing. The gel is allowed to polymerize for 1 h with comb in place. Electrophoresis is run in 1 × TAE buffer for 1200 V h. The gels are then stained with ethidium bromide (0.5 μg mL^{-1}) for 20 min, destained in water for 20 min and then photographed.

3.4   PCR Contamination and Avoiding False Positives

The most significant problem plaguing PCR is the contamination of reagents and pipettes with previously amplified material. The extreme sensitivity of the PCR reaction is complicated by the occurrence of false positives usually due to carry-over of the greater than 10^6 copies of amplified product. Through aerosol, the amplified DNA fragments can be inadvertently introduced into the next PCR reaction. Therefore, meticulous attention is devoted to the physical separation of pre-PCR and post-PCR manipulations. To prevent carry-over, pre-PCR samples are prepared in a separate room or hood. A separate set of positive displacement pipettes are used. We divide reagents into aliquots to minimize the number of repeated samplings necessary. Standard autoclaving of reagents or U.V. irradiation has little use in preventing contamination problems [31].

Non-specific amplification can be reduced by increasing the annealing temperature, using the 'hot start' technique or by the addition of 1–5% formamide [29]. If the 3′ends of each primer contain only two complementary bases which allow for hybridization at a reduced temperature, the formation of primer dimers is possible. This allows for primer dimer concatamers to form in the PCR reaction which yields products of high molecular weight. The addition of template DNA minimizes the problem, however, primer dimer concatamer formation can be a source of false positives [13]. Thus, negative controls (PCR reaction without DNA) should always be included with each experiment.

3.5   Reproducibility and Artifactual Variation

Due to the low stringency of the primer annealing temperature, there are many factors which influence RAPD patterns. Among the most prominent are DNA quality and concentration, Mg^{++} molarity, primer sequence and temperature profile of thermocycler used. Contaminating RNA in many DNA preparations hinders accurate DNA measurements but this can be obviated by fluorescence measurements. Too little DNA gives a variable banding pattern with additional bands appearing at a higher DNA concentration. At an even higher DNA concentration, non-specific amplification or smearing results [18]. Artifactual bands appear when the Mg^{++} concentration is altered [32]. Because of the base composition of selected primer, each oligonucleotide may require optimization with respect to the amplification conditions. Different thermocycler brand names show different overall temperature profiles that can contribute to artifactual bands [33].

Despite some reservations of RAPD analysis expressed by a few researchers, the RAPD bands are reproducible when using a high level of standardization and internal controls . The bands are scored confidently and show by Mendelian inheritance to be factual. This has

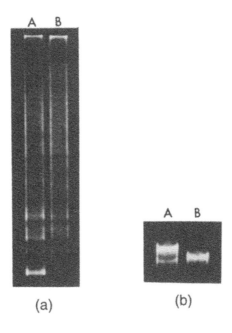

(a)   (b)

Figure 3.   RAPD marker polymorphic phenotype.
(a) Two different clinical isolates (A and B) of Haemophilus somnus show a polymorphic band at the position of the arrow. This corresponds to a phenotype of the presence or absence of a band. (b) Homoduplex mobility shifts are shown on DGGE gels. Two wheat cultivars, A and B, show a RAPD fragment polymorphism. The partial melting of the RAPD fragment A causes a retardation in mobility. Segregation analysis of an F6 population confirms a single genetic locus.

been demonstrated by many investigators [3,11,17]. In our labs, RAPD bands are reproducible in bacteria [13, Figure 3A], protozoa [14] and inheritable for over six generations in the complex, hexaploid wheat genome (unpublished results, Figure 3B). Intense, well resolved bands demonstrate a true genetic polymorphism corroborated by analysis of a segregating population. However, fainter bands are more likely to be scored incorrectly with an error rate in the 2–7% range. These should be ignored in applications requiring a high level of accuracy such as diagnostic testing. Denaturing gradient gel electrophoresis of RAPD bands has a far superior resolving power (Figure 4). By screening a sufficient number of primers and replicate analysis of independent isolated DNA

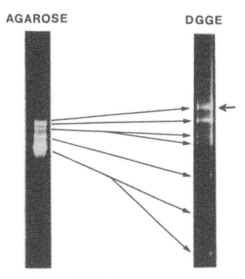

Figure 4. Comparison of RAPD fragments on agarose and DGGE gels. Aliquots from a single PCR-RAPD reaction were run on 1.4% agarose for 240 V h and on 20–60% denaturant gradient gels for 1200 V h Wheat DNA (RL6043) was selectively amplified by UBC primer 5′-TTCCATGCCT-3′. Single bands on agarose are resolved into doublets on DGGE gels. The arrow indicates a band size of about 1.0 kb.

samples, genomic fingerprinting by RAPD analysis has proven to be a valuable analytical tool.

3.6   Construction of Designer or SCAR Primers

For applications requiring high throughput, high level of reliability and universal use, the RAPD polymorphism is converted to a second PCR-based DNA marker. By cloning and sequencing polymorphic bands, SCARs (sequence characterized amplified region) yield information for species specific PCR primers. The sequence of the primers is derived from the termini of the RAPD band (Figure 5). With an increased annealing temperature, these designer primers amplify a single band thus eliminating the multiple banding pattern of RAPDs. The increased specificity of these primers allows their use in related species. In addition, the primers can eliminate the need for post-amplification gel electrophoresis because only a single product is

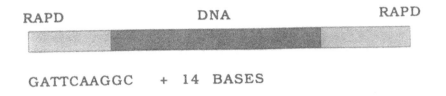

GATTCAAGGC   + 14 BASES

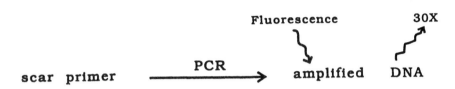

Figure 5.   Construction of <designer> or SCAR primers.
The polymorphic band between the two DNA samples is excised from the
gel and the DNA purified. Standard cloning techniques are used to ligate
the DNA into the TA cloning vector pCRII. The insert DNA is sequenced
using the cloning vector's forward and reverse primers. An arbitrary RAPD
primer's sequence is illustrated. This sequence plus 14 bases of the band's
sequence will yield sequence information for the synthesis of species
specific SCAR primers (24-mers). These primers are now used in the PCR
reaction at a specific annealing temperature. At this temperature,
amplification or no amplification will result depending upon the matching
or mismatching of the primer sequence to template DNA. Polymorphism is
measured by fluorescence (bisbenzamide dye) of amplified DNA.

amplified. In keeping with high throughput applications, the DNA of
the PCR product can be measured by fluorescence using a microplate
fluorometer.

ACKNOWLEDGMENTS

We like to thank Mike Wolf for his expert technical assistance on DGGE
gels and Reg Sims for preparing the graphics. We are grateful for the many
helpful discussions with Drs. W. Kim, G. Penner, and N. Howes.

REFERENCES

[1] Saiki, R.K., Scharf, S., Falonna, F., Mullis, K.B., Horn, G.T. and Arnheim, N. Enzymatic amplification of beta-globin genomic sequences and restriction site analysis for diagnosis of sickle cell anemia. Science, 230 (1985) 1350–1354.

[2] Mullis, K., Faloona, F., Scharf, S., Saiki, R., Horn, G. and Erlich, H. Specific enzymatic amplification of DNA in vitro: the polymerase chain reaction. Cold Spring Harbor Symp. Quant. Biol., 51 (1986) 263–273.

[3] Reiter, R.S., Williams, J., Feldmann, K.A., Rafalski, J., Tingey, S. and Scolnik, P. Global and local genome mapping in *Aradopsis thaliana* by using recombinant inbred lines and random amplified polymorphic DNAs. Proc. Natl. Acad. Sci. USA., 89 (1992) 1477–1481.

[4] Love, J., Knight, A.M., McAleer, M.A. and Todd, J.A. Towards construction of a high resolution map of the mouse genome using PCR-analysed microsatellites. Nucleic Acids Res., 18 (1990) 4123–4130.

[5] Kasi, K., Nakamura, Y. and White, R. Amplification of a variable number of tandem repeats (VNTR) locus (pMCT118) by the polymerase chain reaction (PCR) and its application to forensic science. J. Forensic Sci., 35 (1990) 1196–1200.

[6] Brisson-Noel, A., Gicquel, B., Lecossier, D., Levy-Frébault, V., Nassif, X. and Hance, A.J. Rapid diagnosis of tuberculosis by amplification of mycobacteria DNA in clinical samples. Lancet, 2 (1989) 1069–1071.

[7] Gasparini, P., Novelli, G., Savoia, A., Dallapiccola, B. and Pignatti, P.F. First-trimester prenatal diagnosis of cystic fibrosis using the polymerase chain reaction: report of eight cases. Prenat. Diagn., 9 (1989) 349–355.

[8] Southern, E.M. Detection of specific sequences among DNA fragments separated by gel electrophoresis. J. Mol. Biol., 98 (1975) 503–517.

[9] Tragoonrung, S., Kanazin, V., Hayes, P. and Blake, T.K. Sequence-tagged-site facilitated PCR for barley genome mapping. Theor. Appl. Genet., 84 (1992) 1002–1008.

[10] Botstein, D., White, R.L., Skolnick, M. and Davis, R.W. Construction of a genetic map in man using restriction fragment length polymorphisms. Am. J. Hum. Genet., 32 (1980) 314–331.

[11] Williams, J.G.K., Kubelik, A.R., Livak, K.J., Rafalski, J.A. and Tingey, S.V. DNA polymorphisms amplified by arbitrary primers are useful as genetic markers. Nucleic Acids Res., 18 (1990) 6531–6535.

[12] Welsh, J. and McClelland, M. Fingerprinting genomes using PCR with arbitrary primers. Nucleic Acids Res., 18 (1990) 7213–7218.

[13] Myers, L.E., Silva, S.V.P.S., Procunier, J.D. and Little, P.B. Genomic fingerprinting of 'Haemophilus somnus' isolates by using a random-amplified polymorphic DNA assay. J. Clin. Microbiol., 31 (1993) 512–517.

[14] Procunier, J.D., Fernando, M.A. and Barta, J.R. Species and strain differentiation of *Eimeria* spp. of the domestic fowl using DNA polymorphisms amplified by arbitrary primers (RAPD PCR). Parasit. Res., 79 (1993) 98–102.

[15] Goodwin, P.H. and Annis, S.L. Rapid identification of genetic variation and pathotype of *Leptosphaeria maculans* by random amplified polymorphic DNA assay. Appl. Environ. Microbiol., 57 (1991) 2482–2486.

[16] Wilde, J., Waugh, R. and Powell, W. Genetic fingerprinting of *Theobroma* clones using randomly amplified polymorphic DNA markers. Theor. Appl. Genet., 83 (1992) 8?1–877.

[17]   Echt, C.S., Erdahl, L.A. and McCoy, T.J. Genetic segregation of random amplified polymorphic DNA in diploid alfalfa. Genome, 35 (1992) 84–87.

[18]   Devos, K.M. and Gale, M.D. The use of random amplified polymorphic DNA markers in wheat. Theor. Appl. Genet., 84 (1992) 567–572.

[19]   Fischer, S.G. and Lerman, L.S. DNA fragments differing by single base-pair substitutions are separated in denaturing gradient gels: correspondence with melting theory. Proc. Natl. Acad. Sci. U.S.A., 80 (1983) 1579–1583.

[20]   Myers, R.M., Maniatis, T. and Lerman, L.S. Detection and localization of single base changes by denaturing gradient gel electrophoresis. Methods Enzymol., 155 (1987) 501–527.

[21]   Deweikat, I., Mackenzie, S., Levy, M. and Ohm, H. Pedigree assessment using RAPD-DGGE in cereal crop species. Theor. Appl.Genet., 85 (1993) 497–505.

[22]   Shaw, C.R. Electrophoretic variations in enzymes. Science, 149 (1965) 936–943.

[23]   Howes, N.K., Kovacs, M.I., Leislie, D., Dawood, M.R. and Bushuk, W. Screening of durum wheats for pasta-making quality with monoclonal antibodies for gliadin 45. Genome, 32 (1989) 1096–1099.

[24]   Murray, M.G. and Thompson, W.F. Rapid isolation of high molecular weight plant DNA. Nucleic Acids Res., 8 (1980) 4321–4326.

[25]   Pikaart, M.J. and Villeponteau, B. Suppression of PCR amplification by high levels of RNA. BioTechniques, 14 (1993) 24–25.

[26]   Sambrook, J., Fritsch, E.F. and Maniatis, T. *Molecular Cloning: a Laboratory Manual*, 2nd ed., Cold Spring Harbor Laboratory Press, Cold Spring Harbor, N.Y., 1989, p. 468.

[27]   Edwards, K., Johnstone, C. and Thompson, C. A simple and rapid method for the preparation of plant genomic DNA for PCR analysis. Nucleic Acids Res., 19 (1991) 134.

[28]   Jhingan, A.K. A novel technology for DNA isolation. Methods in Mol. Cell. Biol., 3 (1992) 15–22.

[29]   Sarkar, G., Kapelner, S. and Sommer, S.S. Formamide can dramatically improve the specificity of PCR. Nucleic Acids Res., 18 (1990) 24.

[30]   Chou, Q., Russell, M., Birch, D., Raymond, J. and Bloch, W. Prevention of pre-PCR mis-priming and primer dimerization improves low-copy-number amplifications. Nucleic Acids Res., 7 (1992) 1717–1723.

[31]   Dwyer, D.E. and Saksena, N. Failure of ultra-violet irradiation and autoclaving to eliminate PCR contamination. Molecular and Cellular Probes, 6 (1992) 87–88.

[32]   Ellsworth, D., Rittenhouse, D. and Honeycutt, R.L. Artifactual variation in randomly amplified polymorphic DNA banding patterns. BioTechniques, 14 (1993) 214–217.

[33]   Penner, G.A., Bush, A., Wise, R., Kim, W., Domier, L., *et al.* Reproducibility of random amplified polymorphic DNA (RAPD) analysis among laboratories. PCR Methods and Applications, 2 (1993) 341–345.

Chapter 19

PCR WALKING

Michael G. Burdon and John J. Willoughby

OUTLINE

Methods in Gene Technology, Volume 2, pages 337–348
Copyright © 1994 JAI Press Ltd
All rights of reproduction in any form reserved
ISBN: 1-55938-264-3

1. CHROMOSOME WALKING

The principle of chromosome walking is to allow the isolation by cloning (or by amplification) of DNA sequences which lie adjacent to a segment that has already been isolated, or whose sequence is known. The traditional approach to this problem has been to isolate clones that have partial sequence overlaps with (and hence hybridise with) the starting segment. This process can then be repeated, giving rise to a collection of overlapping clones, or 'contig'. In practice, walks of any length using this method tend to be very laborious, not least because it has to be ensured at each stage that the sequence used to identify the adjacent clone is a single-copy sequence and this may therefore involve a sub-cloning step before the next stage of the walk.

A more athletic way of proceeding along a chromosome, known as chromosome jumping, has been devised. The principle of this method involves producing very long restriction fragments by partial digestion, followed by circularization by self-ligation at very low DNA concentration. This in effect ligates together two DNA sequences which originally lay far apart on the chromosome. By cloning the segment containing this junction, a jump has been made from one sequence to the other. This allows much larger individual steps to be taken than the more pedestrian method outlined above. In addition, it has the great advantage that it can simply step over sequences which, for whatever reason, prove difficult to clone and would thus bring a conventional chromosome walk to a halt. It is not technically simple to carry out, but it has led to some spectacular advances, notably in the collaboration that successfully cloned the gene for cystic fibrosis [1].

2. PCR WALKING

A conventional PCR reaction amplifies the DNA lying between two points on the chromosome which are represented by the two oligonucleotide primers. PCR walking methods involve the modification or subversion of this technique with a view to amplifying flanking DNA lying adjacent to one (or both) of the primers but lying outside the region between them. In effect there are two basic strategies that can be followed in order to pursue this aim.

One, exemplified by the 'inverse PCR' procedure, rearranges the DNA in such a way that sequences that originally lay outside the region between the amplification primers now reside between them. This method will be described below. The other strategy involves placing a known sequence (often another synthetic oligonucleotide, generally by ligation) at a point on the DNA some way from the amplification primer sites and then carrying out a PCR using this new sequence and one of the original primers to amplify the DNA they span. A number of different variants of this strategy have been published and some of them will be outlined below. The method developed in our laboratory, which will be described in some detail, was originally designed to allow PCR to be carried out in circumstances where only enough sequence information was available to design one gene-specific primer, for example when only a short sequence of N-terminal amino acids from a purified protein is known.

The adaptation of the polymerase chain reaction (PCR) to chromosome walking has the immediate advantage of the relative simplicity and speed of the operations involved. However, the inherent limitation of the size of DNA segment that can be amplified efficiently to a few kilobases at most, means that it is restricted to relatively short walks. This limitation would always have to be taken into account when making the decision as to whether to use a PCR-based walking method or a conventional one, setting it against the practical advantages of the PCR methods. As the majority of PCR walking methods require modification of the DNA template by digestion with a restriction enzyme, success may sometimes be aided by knowledge of the presence of appropriate restriction sites lying at a useful distance from the starting point of the walk. As short template fragments of under 1 kb are amplified so efficiently, it may be necessary to carry out a size-selection step before amplification if, for example, a partial digest with a four-base cutter is used as template and a longer walk is required. Thus the availability of information about suitable restriction sites may be an important factor in the decision as to whether to use a PCR walking method or not.

3. INVERSE PCR

As indicated above, the inverse PCR technique rearranges sequences flanking the 'core' sequence (the sequence lying between the two

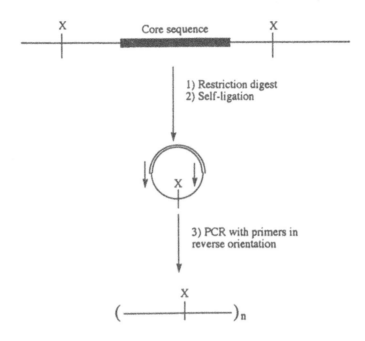

Figure 1. Inverse PCR. The DNA is first digested with a restriction enzyme (Step 1) and then self-ligated at low DNA concentration to give circular products (Step 2). PCR amplification is then carried out with primers in the reverse orientations as described in the text (Step 3).

primers) in such a way that the flanking sequences also become targets for amplification. It was devised more or less simultaneously in more than one laboratory [2,3] and has been discussed in some detail by Ochman etal. [4]. Figure 1 illustrates the three stages involved in inverse PCR. The first stage involves the use of a restriction enzyme (or possibly two different enzymes that leave compatible ends, such as BamHI and BglII, for example). The enzyme(s) must be chosen so that no cleavage within the core sequence takes place. Thus a fragment is produced which contains an intact core sequence with a certain amount of flanking DNA lying on either side of it.

The second step is a ligation done at high dilution in order to promote the formation of monomeric circles. Note the importance of using dilute DNA for ligation, in contrast to the high DNA concentrations usually employed in ligation reactions for cloning purposes. Failure to recognise this is the most common cause of failure of inverse PCR. The DNA concentration in the ligation step should not exceed 0.5 μg/mL. (This intramolecular ligation reaction is also a key step in the 'chromosome jumping' procedure mentioned above).

The circular DNA molecule that results from this ligation can be used as a target for PCR in two ways. One would simply be to use the two initial primers, but as this would only amplify the core sequence nothing would have been achieved. But, as shown in Figure 1, if PCR is carried out using primers that are complementary to the original primers (and thus point in the opposite direction), the sequence that will be amplified will be the now contiguous flanking sequences, thus achieving the original aim.

Inverse PCR has the disadvantage that it amplifies sequences on both sides of the core sequence together, and bidirectional walking clearly has problems when it comes to knowing where one has arrived. The ideal situation for a directed walk would be to use a restriction enzyme that cleaves very close to one side of the core sequence and some distance from the other, thus only allowing a significant walk away from the latter side. Treatment of the inverse PCR product with the original restriction enzyme will, of course, separate the two parts and their physical separation could also be used to walk in either direction at will. This option might not be available if different enzymes with compatible cohesive (or blunt) ends had been used.

Inverse PCR can also be performed on sequences that have been cloned into some vector or other. The known restriction sites on the vector may then in principle be used to generate the target fragment for inverse PCR. This could be particularly useful if the fragment is cloned into a multiple cloning site (polylinker) such as are found in many commonly used vectors.

4. OLIGONUCLEOTIDE-MEDIATED PCR WALKING

The basic requirement for this approach is that one should know enough sequence information in order to be able to design one oligonucleotide primer. The sequence information needed is thus less than that needed for a standard PCR reaction. Enough data for a

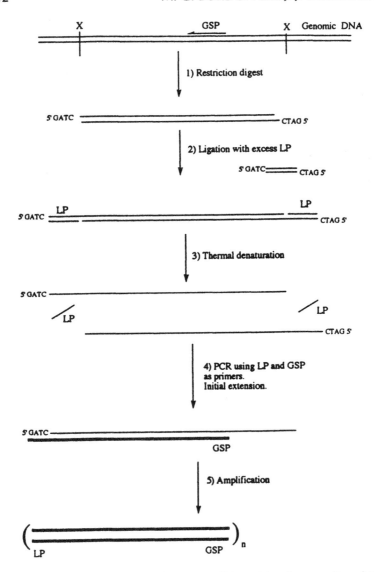

Figure 2. Ligation-mediated PCR walking. Ligation-mediated PCR walking. After the initial restriction digest (Step 1), ligation is carried out with a large molar excess of the ligated primer (LP; Step 2). As the DNA is not phosphorylated, thermal denaturation (Step 3) leaves the LP attached at only the 5′-ends of the DNA strands. Step 4 is then the extension of the GSP (gene-specific primer) during the first PCR cycle. Subsequent PCR cycles then exponentially amplify the sequence flanked by the GSP and the LP.

single primer can come, as noted above, from a short stretch of amino acids obtained by N-terminal sequencing of a polypeptide. In order to carry out the PCR, and thereby to perform a walk away from this initial primer sequence, a second primer sequence has to be planted at an appropriate distance from the initial one, generally by a ligation reaction. A standard PCR can now be carried out using this new primer and the initial one. Several variants of this method have been published [5-7] and the one described here was developed in our laboratory during a study of the urease gene of Ureaplasma urealyticum [5]. A urease-specific primer was designed on the basis of the N-terminal amino acid sequence obtained by direct sequencing of one of the subunits of the urease enzyme. This primer is labelled GSP (gene-specific primer) in Figure 2, which shows the steps involved in this method of PCR walking.

The first step is the cleavage of the DNA with a restriction enzyme (or enzymes) that give cohesive ends compatible with the primer sequence to be attached by ligation (referred to as LP, for ligated primer). Enzymes giving 5´-GATC overhangs were used, as there are several readily available ones. BamHI and BclI turned out to be particularly useful in our study. The LP used is self-complementary and forms a symmetrical duplex which contains restriction sites for possible use in later steps, such as cloning. When it self-anneals, it gives a symmetrical duplex with 5´-GATC overhangs, thus:

```
                        HindIII
5´- G A T C C G A A T T C A A G C T T G A A T T C G
            G C T T A A G T T C G A A C T T A A G C C T A G - 5´
            EcoRI                       EcoRI
```

An important feature of this oligonucleotide is that it is not phosphorylated at its 5´-end; this is necessary to ensure that other, non-specific fragments are not amplified due to the ligation of LP at both ends. Thus when the LP oligonucleotide is ligated to the cohesive ends left by the restriction digest of the starting DNA, only one phosphodiester bond can be formed. Following the first denaturation step of the PCR, therefore, each single-stranded restriction fragment will only have the LP attached at its 5´-end. The result is that only that fragment which can anneal to the gene-specific primer and carries the LP at its 5´-end will be amplified. All other fragments carry LP only at one end and will therefore not be amplified.

This process is illustrated in Figure 2 and the amplified fragment constitutes a walk from the gene-specific primer to the restriction site to which the LP has been ligated. The direction of the walk is clearly determined by the orientation of the gene-specific primer, being in the 5´→3´ direction. A walk in the opposite direction could be performed by using the complementary primer with the opposite orientation.

5. POTENTIAL PROBLEMS

Although this method of carrying out a short walk along genomic DNA has proved relatively straightforward (we were able to amplify and sequence the urease gene from Ureaplasma urealyticum by making two steps using this method), there are two potential problems that must be borne in mind. The first is the question of specificity. The high degree of specificity of the normal PCR relies on the fact that both primers are unique sequences within the genome so that, even if one primer finds another sequence showing partial homology with it, the chance that the other primer will also find such a sequence within a short distance of it is extremely small. However, everybody who uses PCR knows that unexplained bands, presumably due to some sort of non-specific amplification, can from time to time occur and further steps have to be taken to ascertain which is the genuine PCR product. Any PCR procedure that uses only one specific primer sequence, as is the case with this method of PCR walking, is clearly more susceptible to problems of specificity.

As can be seen in Figure 2, the first reaction in the PCR is the extension of the GSP alone. Although the LP is symmetrical, and can therefore potentially anneal to either strand, only one of these targets (at the 5´-end) is present in the product of step 3 (Figure 2). The 3´-end target site, which is required for PCR amplification, is not present initially, but is formed as a result of the first PCR round with the GSP. Therefore, only after the initial extension of the GSP, followed by the next denaturation step, will the LP bind in the correct orientation and the normal PCR amplification can then proceed.

However, the problem of reduced specificity arises from the fact that, if the GSP finds another partially homologous sequence elsewhere in the genome and is able to prime DNA synthesis with any significant efficiency, that initial product will then be amplified effectively in subsequent PCR cycles. With this procedure, it is therefore even more important than when carrying out conventional PCR reactions to be aware of the possibility of obtaining 'false' PCR products due to non-specific priming. This problem is clearly more likely to be troublesome when dealing with more complex genomes. In addition to using the PCR primers at the highest feasible annealing temperature, in order to minimise any such reduced specificity, a 'nested' approach may be useful in overcoming this problem. The product of the initial PCR walk, from the known gene-specific primer to the non-specific primer, is then used as the template for a second PCR reaction in which a second gene-specific primer replaces the original one. Clearly such a procedure is more time consuming, but it may pay dividends in terms of the increased specificity that can be obtained.

The second problem can also give rise to bands of unexpected sizes, but is less serious in that they are genuine bands in the sense that DNA sequences lying adjacent to the GSP have been amplified. The problem arises from the fact that some restriction enzymes cut DNA, albeit rarely, at sequences that differ from their standard recognition sequences. The so-called 'star' activity of EcoRI (and to a lesser extent BamHI) is a case in point. In a normal restriction digest a small percentage of such illegitimate cleavage may not even be detected but, when coupled with the PCR, it may become magnified out of all proportion. This could particularly be the case if the incorrect PCR walking product was much shorter, and was amplified more efficiently, than the correct one. As stated above, the walking product is still a genuine one but, as it gives rise to a product of unexpected length, it could easily be assumed to be due to non-specific priming.

It is not easy to give general advice on the problem of non-specific, or unexpected, bands following PCR walking by this method. Each particular situation has to be treated individually depending on the amount of information available in terms of DNA sequences and the location of neighbouring restriction sites. Confirmation that a PCR product is the correct one may involve simply a restriction digest if the restriction map is known. Alternatively hybridization to a second gene-specific oligonucleotide could be used if enough sequence information for its synthesis is available.

6.   OTHER METHODS OF PCR WALKING

As mentioned above, a number of detailed schemes have been devised for carrying out PCR walking and several of these are variants in one way or another of the oligonucleotide-mediated method already described. This survey will be concluded by a brief description of three such methods, each one of which employs rather different principles to achieve a similar objective.

This first method is that of Rosenthal and Jones [8] who, in common with our method, ligate an 'oligo-cassette' with a cohesive end onto genomic restriction fragments. The PCR procedure is, however, divided into two steps, the first of which involves the linear amplification of the sequence using the specific primer (biotin labelled) alone. The products from this step are isolated at this stage by binding to streptavidin-coated magnetic beads, before proceeding to a conventional PCR step using both primers.

An ingenious system for the amplification of unknown flanking DNA has recently been published by Jones and Winistorfer [9]. Once again an oligonucleotide is ligated to the ends of the restriction fragments, but in this method the ligated sequence corresponds to a sequence known to lie within the target DNA. Following denaturation and intra-strand reannealing, therefore, the DNA forms a 'panhandle' structure in which the 3´-end of the ligated oligonucleotide is able to prime synthesis into the unknown flanking region, thus achieving a PCR walk, as this newly synthesized DNA can then be amplified in the conventional way.

A method of PCR walking which employs no ligation step has been investigated by Parker et al. [10]. One known primer, with its 3´-end directed towards the unknown flanking sequences to be amplified is used and PCR reactions are then carried out with other primers chosen essentially at random. Some of these will, by good fortune, have sufficient sequence homology with some sequence present on this target DNA to prime the PCR, thus carrying out a PCR walk in a single step. This method clearly has the advantage that no additional steps are required, with their associated possibilities of error or uncertainty, but it is clearly only a practical approach in laboratories where a large number of PCR primers have already been synthesized for other work. With this limitation in mind, it certainly seems to give more consistent results than one might intuitively have expected.

This brief survey of other methods for PCR walking is by no means exhaustive and other variants are being devised and published as time goes on, often with specific situations in mind. PCR walking is clearly not a method for making giant strides along a chromosome, but as a means of getting relatively easy access to nearby flanking sequences it shows a great deal of promise.

ACHNOWLEDGMENTS

The work described in this Chapter was funded by a grant from the Wellcome Trust. We are indebted to Margaret Wilson for typing the manuscript and to Tony Butler for help with the preparation of the figures.

REFERENCES

[1] Rommens, J.M., Iannuzzi, M.C., Kerem, B., Drumm, M.L., Melmer, G., Dean, M. et al., Identification of the cystic fibrosis gene: chromosome walking and jumping. Science, 245 (1989) 1059–1065.

[2] Ochman, H., Gerber, A.S. and Hartl, D.L., Genetic applications of an inverse polymerase chain reaction. Genetics, 120 (1988) 621–623.

[3] Triglia, T., Peterson, M.G. and Kemp, D.J., A procedure for in vitro amplification of DNA segments that lie outside the boundaries of known sequences. Nucleic Acids Res., 16 (1988) 8186–8186.

[4] Ochman, H., Ajioka, J.W., Garza, D. and Hartl, D.L. In Erlich, H.A. (ed.), PCR Technology: Principles and Applications for DNA Amplification, Stockton Press, New York, 1989, pp. 105–111.

[5] Willoughby, J.J., Russell, W.C., Thirkell, D. and Burdon, M.G., Isolation and detection of urease genes in Ureaplasma urealyticum. Infection and Immunity, 59 (1991) 2463–2469.

[6] Kalman, M., Kalman, E.T. and Cashel, M., Polymerase chain reaction (PCR) amplification with a single specific primer. Biochem. Biophys. Res. Comm., 167 (1990) 504–506.

[7] Shyamala, V. and Ferro-Luzzi Ames, G., Genome walking by single-specific-primer polymerase chain reaction: SSP-PCR. Gene, 84 (1989) 1–8.

[8] Rosenthal, A. and Jones, D.S.C., Genomic walking and sequencing by oligo-cassette mediated polymerase chain reaction. Nucleic Acids Res., 18 (1990) 3095–3096.

[9] Jones, D.H. and Winistorfer, S.C., Sequence specific generation of a DNA panhandle permits PCR amplification of unknown flanking DNA. Nucleic Acids Res., 20 (1992) 595–600.

[10] Parker, J.D., Rabinovitch, P.S. and Burmer, G.C., Targeted gene walking polymerase chain reaction. Nucleic Acids Res., 19 (1991) 3055–3060.

INDEX

Methods in Gene Technology

Edited by **Jeremy W. Dale** and **Peter G. Sanders**,
*Molecular Biology Group, Department of Microbiology,
University of Surrey, England*

Gene probes, whether RNA or DNA, have played a central role in the rapid development of molecular biology. The wide variety of applications is matched by a considerable diversity in the methods used for generating probes. A complete account of the available methods and the applications would be an impossible task. Instead the Editors of this new, continuing series have attempted a compromise by combining a selection of newer procedures with a review of the most important established methods, together with some examples of the ways in which gene probes can be applied. By doing so, the aim is to provide not only an introductory manual that will be of benefit to newcomers to the field, but also to broaden the horizons of existing researchers.

Volume 1, 1991, 300 pp.
ISBN 1-55938-263-5

Among the themes explored in this inaugural volume of *Methods in Gene Technology* are the generation of labelled probes; hybridisation; and the uses and methods of detection of isotopic and non-isotopic probes.

J A I
P R E S S

Advances in Gene Technology

Edited by **Peter J. Greenaway,** *Division of Pathology, PHLS Centre for Applied Microbiology and Research, Porton Down, England*

The techniques associated with genetic manipulation are now being applied to a large number of research and industrial problems. This new series aims to keep abreast of developments in the general fields of molecular biology and biotechnology by providing state-of-the-art reviews from scientists with specialist experience in particular areas.

Volume 1, 1990, 274 pp.
ISBN 1-55938-204-X

The first volume brings together articles on eukaryotic gene expression, vector development, DNA separation techniques, the uses of DNA hybridization, transposon mutagenesis and self-splicing mechanisms. These articles are succinct and comprehensive, and indicate the scope of the subject matter covered in this and subsequent volumes. They will be of considerable interest to graduate students, research workers and all those who wish to be kept aware of progress in these fast-moving and rapidly expanding areas of research.

Volume 2, 1991, 250 pp.
ISBN 1-55938-268-6

Microbiology and Research, Porton Down. **Structure, Organization and Function of Transfer RNA genes from the Cellular Slime and Mold** *Dictyostelium discoldeum,* R. Marschalek and T. Dingermann, University of Erlangen-Nurmberg. **Optimizing Post-Translational Steps of Gene Expression in** *Escherichia coli, J.E.G.* McCarthy, GBF, Braunschweig. **Resistance of Herpesvirus and HIV to Antiviral Drugs,** *G. Darby and B.A. Larder, Wellcome Research Laboratories, Beckenham.* **The Molecular Genetics of Host Specificity in Plant Pathogenic Bacteria,** *A. Vivian, Bristol Polytechnic.* **The Cloning and Structure of Genes from the Autotrophic Biomining Bacterium,** *Thibacillus ferrooxidans,* D.E. Rawlings, D.R. Woods and N.P. Mjoli, University of Cape Town, Rondebosch. **Direct Sequencing of RNA Using Dideoxynucleotides,** *M. Kaartinen, M.L. Solin and O. Makela, University of Helsinki.* **Extrachromosomal Elements of the Genus** *Chimydia, M. Lusher, C.C .Storey and S.J. Richmond, University of Manchester.* **The Cytomegalovirus Major Immediate Early Promoter and Its Use in Eukaryotic Expression Systems,** *G.W.G. Wilkinson and A. Akrigg, PHLS Centre for Applied Microbiology and Research, Porton Down.*

Volume 3, In preparation, Spring 1994
ISBN 1-55938-475-1

Advances in DNA Sequence Specific Agents

Edited by **Laurence H. Hurley,** *Drug Dynamics Institute, The University of Texas at Austin*

Sequence recognition of DNA can be achieved by DNA binding proteins and small molecular weight ligands. The molecular interactions which lead to sequence recognition are of considerable importance in chemistry and biology. This series entitled *Advances in DNA Sequence Specific Agents* will examine the techniques used to study DNA sequence recognition and the interactions between DNA and protein and small molecular weight molecules which lead to sequence recognition.

Volume 1, 1992, 347 pp.
ISBN 1-55938-165-5

CONTENTS: Introduction to Series: An Editor's Foreword, *Albert Padwa.* **Preface. PART I: METHODS USED TO EVALUATE SEQUENCE SPECIFICITY OF DNA REACTIVE COMPOUNDS. Application of Equilibrium Binding Methods for Determination of DNA Sequence Specificity,** *Jonathan B. Chaires, University of Mississippi, Medical Center.* **Quantitative Aspects of Dnase 1[+] Footprinting,** *Jim Dabrowiak, Syracuse, New York.* **Use of Circular Dichroism to Probe DNA Structure and Drug Binding to DNA,** *Christoph Zimmer, Jena, and Gerhard Luck, Institut fur Microbiologie und Experimentelle Therapie Jena/Thuringen.* **NMR Analysis of Reversible Nucleic Acid-Small Molecule Complexes,** *W. David Wilson, Ying Li, and James M. Veal, Georgia State University.* **Use of Enzymatic and Chemical Probes to Determine the Effect of Drug Binding on Local DNA Structure,** *Keith R. Fox, University of Southampton,*

JAI PRESS

England. **PART II: SEQUENCE SPECIFICITY OF DRUGS THAT INTERACT WITH DNA IN THE MINOR GROVE. The DNA Sequence Selectivity of CC-1065,** *Martha Warpehoski, Upjohn, Michigan.* **Mitomycin C: DNA Sequence Specificity of a Natural DNA Cross-Linking Agent,** *Maria Tomasz, Hunter College.* **Sequence Specificity of the Pyrrolo(1,4)Benzodiazepines,** *John A. Mountzouris and Laurence Hurley, University of Texas.* **Calicheamicin/Esperamicin,** *George Ellestad and Nada Zein, Lederle, New York.* **DNA Sequence Control Mechanism of Oxidative Deoxyribose Damage by Neocarzinostatin,** *Irvine Goldberg, Harvard University.*

Volume 2, In preparation, Spring 1994
ISBN 1-55938-166-3

Advances in Developmental Biochemistry

Edited by **Paul Wassarman,** *Department of Cell and Developmental Biology, Roche Institute of Molecular Biology,*

Volume 1, 1991, 256 pp.
ISBN 1-55938-347-X

CONTENTS: Introduction. Organelle Assembly and Function in the Amphibian Germinal Vesicle, *Joseph G. Gall, Carnegie Institution.* **DNA Replication and the Role of Transcriptional Elements During Animal Development,** *Melvin L. DePamphilis, Roche Institute of Molecular Biology, New Jersey.* **Transcriptional Regulation During Early** *Drosophila* **Development,** *K. Prakash, Joanne Topol. C.R. Dearolf, and Carl S. Parker, California Institute of Technology, Pasadena.* **Translational Regulation of Maternal Messenger RNA,** *L. Dennis Smith, University of California, Irvine.* **Gut Esterase Expression in the Nematode** *Caenorhabditis Elegans,* *James D. McGhee, University of Calgary.* **Transcriptional Regulation of Crystallin Genes: Cis Elements, Trans-factors and Signal Transduction Systems in the Lens,** *Joram Piatigorsky and Peggy S. Zelenka, National Eye Institute, National Institutes of Health, Maryland.* **Subject Index.**

Volume 2, 1993
ISBN 1-55938-609-6

CONTENTS: Preface, *Paul M. Wassarman, Roche Institute of Molecular Biology.* **Drosophila Homeobox Genes,** *William McGinnis, Yale University.* **Structural and Functional Aspects of Mammalian HOX Genes,** *Denis Duboule, European Molecular Biology Laboratory.* **Developmental Control Genes in Myogenesis of Vertebrates,** *Hans Henning-Arnold, University of Hamburg.* **Mammalian Fertilization: Sperm Receptor Genes and Glycoproteins,** *Paul M. Wassarman, Roche Institute of Molecular Biology.* **The Fertilization Calcium Signal and How It Is Triggered,** *Michael Whitaker, University College London.* **Subject Index.**

Advances in Genome Biology

Edited by **Ram S. Verma,** *Division of Genetics,*
Long Island College Hospital

Volume 1, Unfolding the Genome
1991, 425 pp.
ISBN 1-55938-349-6

**CONTENTS: Genetic Techniques for Mapping and Sequencing the
Genome,** *Ram S. Verma, Long Island College Hospital - SUNY.* **Cloning
Defined Regions of the Genome,** *Bernard Horsthemke, Universitatsklin-
ikum Essen, Germany.* **In Situ Hybridization,** *Paul Szabo, Cornell University
Medical College.* **Southern Blotting,** *K. Stam, Oncogene Science Inc., New
York.* **Northern Blotting,** *M.A. Q. Siddiqui, SUNY Health Science Center,
New York.* **Slot Blot Technique: Principles and Application,** *Geeta
Vasanthkumar, Southern Research Institute, Alabama.* **Dot Blot
Technique,** *Roger Lebo, University of California.* **Polymerase Chain
Reaction Technique: Principles and Application,** *Stephen M. Carleton,
SUNY Health Science Center, New York.* **Pulse-Field Gel Electrophoresis:
Detection of Large DNA Molecules,** *Robert M. Gimmill, Eleanor Roosevel
Institute for Cancer Research, Colorado.* **Detection of Single Base Charge
in Nucleic Acid,** *Richard G.H. Cotton, Royal Children's Hospital, Australia.*
**Identification of Chromosome Specific Satellite DNA from the
Centromere by Biotinylated DNA Probes,** *Matteo Adinolfi, United Medical
and Dental School, London, England.* **Construction and Usage of Linkage
Libraries,** *Alan R. Kimmel, National Institute of Diabetes and Digestive
Kidney Diseases, Maryland.* **Subject Index.**

Volume 2, Morbid Anatomy of the Genome
1993
ISBN 1-55938-583-9

CONTENTS: Diagnosis of Human Genetic Disease, *Jorg Schmidtke,
Medizinische Hochschule, Germany.* **Mitochondrial DNA and Disease,**
A.E. Harding, Institute of Neurology, London. **The Cystic Fibrosis Gene,**
*Michael Wagner, Abteilung Kinderheilkunde Schwerpunkt Neuropadiatrie,
Germany.* **Marfan Syndrome: A. Molecular Biology,** *Brendan Lee, SUNY
Health Science Center, New York.* **Marfan Syndrome: B. Molecular
Pathogenesis,** *Leena Peltonen, Laboratory of Molecular Genetics, Finland.*
Molecular Genetics of the Fragile X, *Grant R. Sutherland, Adelaide
Children's Hospital, Australia.* **Molecular Genetics of Huntington Disease,**
C. Pritchard, Johin Radcliffe Hospital, England. **Gene for Von
Recklinghausen Neurofibromatosis Type 1.,** *Dalal M. Jadayel, The
Haddow Laboratories, London.* **Molecular Biology and Duchenne and
Becker Muscular Dystrophies,** *Jamel Chelly, John Radcliffe Hospital,
England.* **Molecular Genetics of Thalassemia,** *Stephen A. Liebhaber,
University of Pennsylvania.* **Application of Molecular Genetics for
Identity,** *Zvi G. Loewy, Hoffmann-LaRoche Inc., New Jersey.* **Current
Status and Future Directions in Human Gene Therapy,** *Paul Tolstoshev,
Genetic Therapy Inc., Maryland.* **Prelude: Reverse Genetics,** *Ram S.
Verma, Long Island College Hospital.* **Subject Index.**

J A I   P R E S S

Advances in Developmental Biology

Edited by **Paul Wassarman**, *Department of Cell and Developmental Biology, Roche Institute of Molecular Biology*

Volume 1, 1991, 192 pp.
ISBN 1-55938-348-8

Volume 2, 1993
ISBN 1-55938-582-0

JAI PRESS LTD   •   JAI PRESS INC.

(United Kingdom. Europe, Middle East, Africa and the Indian Subcontinent)
The Courtyard
28 High Street, Hampton Hill
Middlesex TW12 1PD, U.K.

Tel: 081-943 9296
Fax: 081-943 9317

(The Americas, Western Hemisphere, Pacific Rim, Australia and New Zealand)
55 Post Road No. 2
Greenwich
Conn. 06836-1678, U.S.A.

Tel: (203) 661 7602
Fax: (203) 661 0792

Printed and bound by CPI Group (UK) Ltd, Croydon, CR0 4YY

03/10/2024

01040436-0007